Animal Angels

Animal Angels

Amazing Acts of Love and Compassion

Stephanie Laland

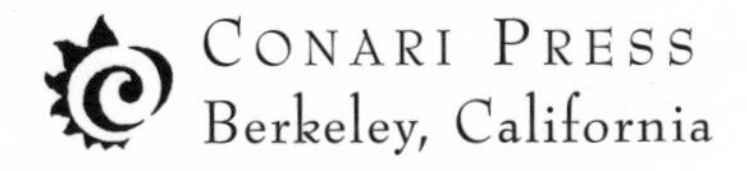

Conari Press books are distributed by Publishers Group West

Cover design by Suzanne Albertson
Cover photo by Kathleen Webster, Tony Stone Images
Interior design by Claudia Smelser

The author wishes to thank:

Jeffrey Armstrong, whose poems appear by permission.
The Ken-L-Ration (now Reward) Awards
The Latham Awards
The Whiskas Vitalife Awards
The William O. Stillman Awards
Wildlife Images of Grants Pass, Oregon
. . . and all the people who wrote to me

Library of Congress Cataloging-in-Publication Data

Laland, Stephanie
Animal Angels ; amazing acts of love & compassion / Stephanie Laland.
p. cm.
Includes bibliographical references.
ISBN: 1-57324-142-3 (trade paper)
1. Altruistic behavior in animals—Anecdotes. 2. Human-animal relationships—Anecdotes. I. Title.
QL775.5.L344 1998 98–36868
591.5—dc21 CIP

Printed in the United States of America on recycled paper
1 2 3 4 5 6 7 8 9 10

To Linda

ANIMAL ANGELS

A WORD FROM TIPPI HEDREN

I HAVE LIVED in the shadow of birds and big cats for many years. I love all creatures big and small, feathered, furred, and finned. What they have taught me is a keen respect for the environment and the natural order of life.

Animals live *with* nature, thereby becoming a *part* of nature. They live with dignity, without cutting a swath of destruction as they pass through the forests, the mountains, and the waterways.

Humans can learn so much from these wonderful creatures. They belong on earth just as much as we do, and their rights need to be protected.

This book is a lovely tribute to all creatures, great and small.

The Sanctity of Life

This is a book about the sacredness of life and the creatures in it. It is called *Animal Angels* because many of the creatures with whom we share planet Earth often act with so much altruism, compassion, and love that if a human being were to act this way he or she would be called a saint. Indeed, many of the animals you are going to read about in these true stories act as courageously, especially in emergency situations, as I would like to think I would.

For generations, animals have been reminding us of the natural love in the universe and delighting us with their playful enjoyment of it. Whether we come home from work tired and stressed to find comfort in a small, mute friend who lets us know that we are the most important being in the world to them, or whether we watch animals in parks or on television and somehow feel more connected to life as it was originally created, we know that animals keep us in touch with an essential innocence, and that if we let that connection die we will either die ourselves or be condemned to living a life that doesn't matter anymore.

If love is a prerequisite for angelhood (that's like sainthood, only friendlier) then animals are well on their way, for they have love in abundance. In researching this book, I have discovered a wealth of wonderfulness in both animals and people. I invite you to share with me the discovery of just how awesomely beautiful, how sweet, how connected, how splendid, our spiritual brothers and sisters in different bodies truly are.

To acknowledge the kindness of animals is not to pretend that many animals are not carnivores. That is a fact of nature. But there is another side to the animal kingdom which we have too long ignored because it has not suited our purposes.

Suppose we were to encounter an alien being who was reporting back to his home planet about our earth and we discovered his notes. If the content of these notes discussed only our wars, gang warfare, politically induced famine, and murders, would we not object? Would we not seek to mention that we have also been blessed with the likes of a Mother Teresa, a Martin Luther King Jr., and a Gandhi?

Our assessment of the animal kingdom has been so biased and one-sided toward the "law of the jungle" that we have dismissed an enormously important aspect of life.

From earthworms to rabbits, from bacteria to chimpanzees, all beings have something precious to give; each has a perfect place in creation. All animals carry within them a special song that makes them essential to the symphony of life. Through our short-sightedness, we have already lost forever some of our fellow members in life's great chorus. Let us awaken our own minds and hearts before we lose any more.

Angels in Feathers & Fur

None is so near the gods as he who shows kindness.
—Seneca

Perhaps the greatest gifts animals offer are those we take for granted. Because a pet is always there for us when we come home, always waiting with unconditional love, we don't think of their devotion as particularly noteworthy. Even heroic acts, such as dogs risking or sacrificing their own lives to save a loved one, are so commonplace that one might be tempted to think, "Oh well, isn't that just what dogs do?"

I am very often touched by hearing the stories told through tears of all that a recently departed animal has meant to someone. But because the love between humans and their companion animals is so common does not make it less precious. To me, it is all the more precious because it is always there.

Fortunately, it is the most important things in life that the Creator has given us most abundantly. Humankind can live without gold and jewels for a lifetime, but only about a month without food; therefore, food is less expensive. We can live without food for a few weeks but without water for only about a week; therefore, water is not as expensive as food. We can live without water for perhaps a week but without air only a few moments; therefore, air is completely free. We can see that that which is most precious—what we cannot live without—is given to us the most freely. But we sometimes fail to see this because we have confused scarcity with value. We often take for granted the love of our companion animals as much as we do the very air we breathe.

Many of the stories in this book are about the exemplary behavior of animals. Many occurred in extreme situations requiring courageous or compassionate action beyond the ordinary. But the animals, especially those with whom we live intimately, are giving us such consistent, daily love that I would like to begin by honoring this less dramatic but profoundly important service.

Even if it is not a dramatic rescue, contact with animals can heal the body and nourish the soul in countless ways. My cats have never rushed into a fire to save me, and I suspect they would not (although there are many accounts of cats who did not simply escape a fire, but leapt upon their guardian companion and pawed at his or her face to awaken them). But my feline companions have enriched my life in many other ways. They have brought me the sweetest, most dependable love. They have cheered my days, ministered to me when I was sick, encouraged me to play when I thought depression was the order of the day, and connected me back again to my own heart when I was becoming estranged from myself.

Are animals angels? Just ask yourself this question: How much less love would there be in our world if the animals were gone? Imagine a world where no dogs frolicked on the beach or filled childhood days with unconditional love, where no horses galloped across the plains, no birdsong filled the morning air, no lions or elephants roamed the African expanses; imagine if our oceans were bereft of dolphins, seals, and whales. What a sad, lonely planet it would be.

The same Power formed the sparrow
That fashioned man the King;
The God of the whole gave a living soul
To furred and to feathered thing.
—*Ella Wheeler Wilcox*

Sometimes we never comprehend how present for us and how important to us an animal is until a terrible crisis arises. Then the soulful devotion and care of our animal friend changes our lives forever, as in this story of a young woman who received an unforgettable lesson in love from her own animal angel.

Elizabeth, a twenty-three-year-old law student, wrote me after reading my book *Peaceful Kingdom: Random Acts of Kindness by Animals*. A month previously, she had received a call from her parents reporting that her twenty-one-year-old brother had just been murdered. She was absolutely devastated and all alone except for the company of her Golden retriever, Paxton.

As she sobbed at the terrible news, Paxton licked her tears away and rested his paw upon her hand. For a time after the funeral, Elizabeth experienced crying spells, insomnia, and general depression. Paxton appeared to be going through the same depression and seemed to be extremely worried about her.

The dog normally slept on the floor beside the bed. But now Paxton knew Elizabeth needed her more, so every night Paxton slept with Elizabeth in her bed. He made sure that some part of his body was always in contact with her, so that she would be reassured by his warm presence at all times.

When she lost her appetite, she was not left alone in this ordeal: Paxton also refused to eat. His behavior mirrored her own as if he were trying to take some part of her sadness into his own body and relieve her of it. By letting her know in every way that he totally loved her and would share her heartbreak with her, Paxton gave Elizabeth the connection and strength she needed to go on.

Eventually the sorrow began to lift. Elizabeth says she could not have even begun to heal without Paxton's love and concern.

"Losing my brother," she wrote, "has certainly changed my life, but through my buddy Paxton I've learned that someday things will again be okay."

Where a human crime had taken love away, a dog with remarkable understanding sought to restore it. Where a human heart was tragically emptied, a dog did his best to fill it again. Sometimes all any angel can do is be living proof of love.

Until one has loved an animal, a part of one's soul remains unawakened.

—Anatole France

To be a stranger in a world in which everyone but you seems to have found a place can be a very lonely experience. In such cases, the adage "a friend in need is a friend indeed" is especially true.

A pet store in Fresno featured a glass-walled bird room containing a large tree whose branches served as bird perches. Various types of parrots had staked out certain branches as their own territory. When a small green parrot was introduced to the aviary, the other birds shunned the young newcomer. When he tried to find a spot on the branches, the established birds vigorously repelled him. The green parrot gingerly tried each branch, but each time he was chased off by the other birds, who flapped their wings, screeched, pecked, and charged at him. When the discouraged little bird came to the last branch, he faced an imposing long-billed red macaw.

The little parrot timidly climbed to the macaw's branch and began to edge slowly toward a bit of empty perch near the larger bird, all the while keeping a wary eye for any sign of hostility.

Suddenly, the red macaw began to raise his wing; the little bird froze. This had signaled the beginning of the territorial screeching and antagonistic display in every case so far. And the macaw was a much bigger bird.

But as he unfurled his large wing, the macaw looked the little parrot in the eye. What the smaller bird saw, or hoped he saw, was not hostility, but friendship. He would have to take the chance. After only a moment's hesitation, the little parrot cautiously edged over and took shelter under the macaw's large and welcoming wing. The macaw then cuddled the little bird close.

This remarkable scene of kindness between strange birds was witnessed by Becky Long, who said, "It was a special event in my life and a moment of such pure sweetness that I will never forget. Before this, I never really gave a lot of thought to the expression "taking someone under your wing.'"

A bird in the hand is a certainty.
But a bird in the bush may sing.

—Bret Harte

Professor Ron Baumgarten wrote me of an act of kindness he had seen: "I observed a group of common sparrows appearing to be trying to help out a fellow sparrow who had something stuck in his beak, chewing gum perhaps. They would take turns trying to pull the foreign material from the victim's beak."

Nothing living should ever be treated with contempt. Whatever it is that lives, a man, a tree, or a bird, should be touched gently, because time is short. Civilization is another word for respect for life.

—Elizabeth Goudge

After speaking on a national radio talk show about the compassion of animals, I received a beautiful letter from a man who had an extraordinary experience with an angelic being named Delphi.

In 1976, this man had been given a beautiful thin-haired Samoyed puppy and named her Delphi. He wrote, "Delphi was the best friend and companion I ever had in my life."

The man went through a painful divorce when his two sons were very young. The youngest boy was afflicted with cystic fibrosis, a fatal genetic disorder. Dealing with his son's medical problems and the emotional scars of the divorce was terribly stressful.

Shortly after the divorce, he learned that Delphi had a cancerous tumor on her left front leg. He had the tumor removed and she seemed to be fine for a seven-year-old dog. But his own life was an emotional wreck as he struggled to recover from the sadness of divorce and handle full-time childcare. Then he faced the loss of his business.

Delphi's tumor returned. Again the vet removed it. Over the years, Delphi's tumor recurred and was removed four or five times. By the time the dog was twelve, the tumor had come back, her health was poor, and she could no longer easily withstand the operation. After discussion with the vet and his sons, the man decided that if the tumor became so painful that Delphi could not walk or live without pain, then she would be put to sleep.

"This was very hard for me," he wrote, "because Delphi was really the most amazing dog—actually *spirit*—that I've ever known. One evening when the kids were in bed and I was watching late-night TV with Delphi at my side, she looked up at me. For an instant it was as if our minds had joined together. I got a clear message from Delphi that she had taken on the cancer to absorb the negative energy in the house as a result of the divorce, the pain the kids were going through over the family breakup, my own pain, and my son's illness. Delphi communicated that she would continue to absorb the negative and destructive energy to spare us as much as possible. She knew this service would eventually claim her life but she wanted us to know that she loved us and that it was really OK.

"This was an awakening to me like never before. This amazing communication was over in flash, but the look in Delphi's eyes was incredible! Over the years I've wondered if it had been only my imagination, a dream, or a fantasy, but honestly, I am forced to accept Delphi's communication as a very real spiritual experience. When I told my kids, they did not think it crazy, they just accepted it as if they had known all the time. To this day, I have no doubt that Delphi willingly took on the cancer to help us out in our time of pain and need."

This man's letter and obvious sincerity touched me deeply.

But did Delphi really communicate this profound spiritual understanding and actually take on cancer to help shoulder the family's pain?

Perhaps the most important things in life are ultimately the most unprovable. The reality of love, goodness, and joy might not be provable in a strict scientific sense, but in real life it is often no more than a look between lovers that seals a lifelong bond or a glance at one's mother that enables a child to have the confidence to perform a difficult feat. A moment of knowing that you are going to do something that the whole world says you cannot do might be the most important moment in your life. Perhaps we cannot prove these moments, but we dare not dismiss them, for life would have none of its richness without them.

> *For fidelity, devotion, love, many a two-legged animal is below the dog and the horse. Happy would it be for thousands of people if they could stand at last before the Judgment Seat and say, I have loved as truly and I have lived as decently as my dog, and yet we call them only brutes.*
>
> —*Henry Ward Beecher*

I always feel a little chastened when I am reminded that while human beings are not completely essential to life on Planet Earth, earthworms are. Without their ability to burrow, which plows and aerates the soil, or their ability to ingest soil and pass it through their bodies, which remineralizes it, the topsoil would cease to exist and we would all die. In other words, earthworms and the world could do quite well without me, but I could not exist without earthworms.

Even things that I personally dislike about earthworms turn out to be perfect. Their slime acts as lubrication to help them pass through the ground effectively. Their shape is ideally suited to their tasks. They have no lungs but breathe through their skin surface. The body of the worm is moist for the same reason the walls of our lungs must be moist.

Although I have yet to come across a story of a family that was rescued from a fire because their pet earthworms got them out, I now see them as diligent, dedicated helpers in God's plan. These humblest of workers have been given the greatest service of keeping the soil fertile so that plants can grow and all creatures can share in the great gift of existence.

Earthworms are the alchemists of the planet. In the Middle Ages, men sought alchemists who were reputed to be able to turn lead into gold. Is that not what we have in the lowly earthworm? He takes the rubbish of the world, the excrement of the creatures, and through the magic of his supple body turns this into soil so full of nutrients that plants can grow and life can flourish. Through their billion little bodies, earthworms take the cast-offs of creation and revitalize them until we have not simply a yellow metal valued by men as wealth, but the true gold of life itself, ready to pour forth and nourish again. The dead and decaying become the basis for all sustenance within the earthworm's tiny body. In this sense, life itself relies upon the most humble creature of all.

> *Love all God's creatures, the animals, the plants. Love everything to perceive the Divine mystery in all.*
>
> —*Fyodor Dostoyevsky*

Diane Livingstone has dedicated a great deal of her life to educating others about the importance of earthworms and the topsoil they supply. One day, at the same moment as she was being introduced as "The Earthworm Lady," a cat brought an earthworm to her and set it at her feet. The cat must have handled it very gently because it was still alive.

J. Allan Boone, the great-grandson of Daniel Boone, wrote that when a white man meets a dangerous snake on a path, the snake will usually try to bite him, whereas if a snake meets an Indian, he will do him no harm. Boone says this is because the snake can sense the thoughts of each person. The white man is usually thinking something like, "Oh no! A snake! I've got to kill it before it kills me!" The Indian is thinking: "Greetings, my beautiful brother. God bless you and may you have a beautiful day." The snake senses the thought vibrations and responds accordingly.

The Circle of Kindness

Animal angels often prove the adage "what goes around, comes around." These stories of the circle of kindness touch us in an especially deep place because they offer a rare glimpse of the underlying linkages between all beings, the great arcs of love that can embrace the world.

Even the tiniest act of kindness can have repercussions in our lives that are completely unforeseen at the time. We are compelled to wonder how much more beautiful and full of love the world would be if each of us acted on the tiny impulses to kindness we get every day. When these opportunities are overlooked, we never get to see what beauty might have grown from them. But when these opportunities are taken, the world is brighter in ways we could never have imagined. Sometimes we even find out later that what seemed like only a tiny event shifted destiny in a way that would forever change our lives.

A young boy bought his pet snake a live rat for food, but for some reason the snake refused to eat it. Perhaps growing fond of the little creature and his antics, the boy decided not to discard the rat. He kept him as a pet and named him Templeton. Templeton soon established himself as a much-loved and appreciated member of the household. Although his cage was always left open, Templeton preferred to sleep in the cage at night (perhaps rather wisely, under the circumstances) and rarely left its safety.

One night someone in the house next door carelessly threw away a jar of paint thinner; it opened and the poison seeped into the ventilation system in the boy's room. As the other members in the household slept unaware, the fumes permeated the room of the sleeping boy, who began slipping into a coma from which he might never return. But Templeton somehow knew what he must do.

The rat left his cage and approached the boy, awakening him by jumping upon his face and batting at him until he managed to pull him back to consciousness. When the boy finally came to, he smelled the deadly odor in his room and got out in time. His life was saved.

This small but resourceful little rat, whom the snake had refused to eat and the boy decided to love, saved the boy from possible brain damage or death.

It can strain one's credulity to conceive how such a simple creature could be capable of the compassion to perform a selfless act of kindness. Such behavior is clearly not based in some automatic, instinctual response. Furthermore, the *London Daily Telegraph* reported a study that found that a rat will press a lever to save another rat from drowning. It seems there is goodness in even the humblest of creatures.

> *Wilt thou draw near the nature of the gods? Draw near them in being merciful. True mercy is nobility's true badge. He who does not restrict harmless conduct to man, but extends it to other animals, most closely approaches divinity.*
>
> —*Porphry*

"What's the matter with you, Cinder?" fireman Lorenzo Abundiz called to his dog, again. Cinder was usually the most enthusiastic of his two Rottweilers, but today as they went for their usual hike in the mountains, Cinder kept dropping behind.

Ever since Lorenzo had received Cinder as a gift for single-handedly pulling a wall on fire off a fellow firemen, saving the man's life, Lorenzo and Cinder had been the best of friends. Lorenzo hadn't wanted another dog; he had just lost one who was very dear to him. But a gift from the grateful man who owed his life to him was a different matter, so he accepted. Later, Lorenzo adopted another Rottweiler, Reeno, who had been abused as a puppy, and the family was complete.

But now Cinder, who had always been the lead dog, lagged behind. Maybe there's something wrong and I should take Cinder to the vet, thought Lorenzo. He loved his dog too much to risk going any further, so the threesome returned home.

There was indeed something wrong—but not with Cinder. Once they were back inside the house, Lorenzo felt an excruciating pain in the left side of his body. He doubled over, staggered to the floor, and began to lose consciousness. This powerful man, who had run unafraid into blazing infernos for most of his adult life, found himself unable to remain conscious long enough to call his wife or even get to a phone. The world blacked out around him.

A moment later he dizzily started to come to: Reeno was desperately licking his face to revive him. Still, he couldn't move. While Reeno licked his face, Cinder reached up to the kitchen counter, knocked down the cordless phone, and brought it over to where Lorenzo lay. The dog, who had never been taught to fetch the phone, had often seen Lorenzo talking into this object. Now, using his nose, Cinder pushed the phone toward Lorenzo's hand.

Lorenzo dialed 911. The paramedics arrived within minutes and applied emergency care on the spot, saving his life. Thanks to Cinder and Reeno, he received treatment in time and fully recovered.

Lorenzo realized that Cinder had somehow been able to sense the onset of his heart attack and that was why the dog had refused to continue the hike. Animals not only have a sixth sense about danger, they seem to have near-miraculous abilities to sense physical problems *before* they manifest.

Since that incident, Lorenzo has met epileptics whose dogs insistently lead them to their bed just before a seizure. Another woman's dog kept trying to bite at a small dark spot on her leg which the woman hadn't noticed. Because of the dog's warning, she had the spot examined; it turned out to be a very deadly cancer. Thanks to her dog, she caught it in time.

Lorenzo eventually went back to work as a fireman. He was later reprimanded for rushing into a burning building to save a dog, at great risk to his own life. "A life is a life," said Lorenzo, after carrying out the grateful animal. "After all, a dog saved *my* life."

Lorenzo was so moved by the intelligence and devotion of Cinder and Reeno that his respect for all life forms has increased and he has become a vegetarian and appeared on many television programs.

THE LITTLE BLACK DOG

I wonder if Christ had a little black dog,
All curly and wooly like mine:
With two long silky ears and a nose, round and wet,
And two eyes, brown and tender, that shine.
I am sure, if He had, that that little black dog
Knew, right from the first, He was God;
That he needed no proof that Christ was divine,
And just worshipped the ground where He trod.
I'm afraid that He hadn't, because I have read

How He prayed in the garden, alone;
For all of His friends and disciples had fled
Even Peter, the one called a stone.
And, oh, I am sure that that little black dog,
With a heart so tender and warm,
Would never have left Him to suffer alone,
But, creeping right under His arm,
Would have licked the dear fingers, in agony clasped,
And, counting all favors but loss,
When they took Him away, would have trotted behind
And followed Him quite to the Cross.

—*Anonymous*

According to a Pawnee legend, a great council is held whenever a human being seeks growth or spiritual counsel, or is in need of an animal friend. The council considers which animal would be best for that person and sends the animal to that person. Sometimes it may be just a brief encounter; while walking you see an animal and feel better for it or receive an insight.

Many times the animal sent is a stricken or helpless creature, to encourage the person to care for it. According to legend, this is why so many people who have taken in stray animals come to feel that the animal they almost didn't rescue became the most precious being they know, and that the day they found him or her was the luckiest day of their lives.

Carol and Ray Steiner were two such folks. They had headaches, high blood pressure, memory loss, and nausea. And now they found they were sleeping nineteen hours a day. They had both been ill and thought that these symptoms were part of some unknown illness.

But on August 22, 1995, their cat began acting strangely, too. Ringo began throwing himself against the front door of the house. Then he went to the back door and threw himself against it as well. When Mrs. Steiner let him out she somehow knew she must follow the cat.

Ringo began digging in some rock landscape covering. Mrs. Steiner immediately smelled a gas leak. The gas company was called and the leak, which had been slowly killing the Steiners and which potentially could have resulted in an explosion and caused many deaths, was sealed.

Said Mrs. Steiner, "To think we almost passed him up because we thought we had enough cats in the house. We adopted him as a stray from a litter found outside my mother's nursing home. We had three cats and didn't want any more. Had we stuck to our convictions, we would be dead today." Ringo, named for his penchant for drumming on the kitchen floor with his paws, was awarded the American Humane Association's Stillman Award for bravery.

> *The Golden Rule must be applied in our relations with the animal world, just as it must be applied in our relations with our fellow-man.*
>
> —*Ralph Waldo Trine*

Forty-five cats and fifteen dogs are born for every man, woman, and child in the United States. Many parents think that by permitting their animal's pregnancy they will show their children the miracle of life. But doing so only adds to the misery of thousands of loving but unwanted cats and dogs who must be routinely euthanized in the animal shelters of every town and city. It is heartbreaking work, but the massive overpopulation of cats and dogs leaves little choice. The only alternative to this ongoing tragedy would be the widespread determination of humans to spay or neuter their animal companions. Please spay and neuter your beloved angel animal friends—it can be the kindest thing you do for the beings you love.

Here is the story of one animal who was very lucky twice.

Although Sheena was a lovely dog, lack of concern about pet overpopulation had created such a surplus of beautiful animals that she was scheduled to be put to sleep. The day before she was to die, an angelic human saw something very special in her, and Sheena was chosen to become a hearing dog.

Hearing dogs must learn to do many tasks, from making their deaf friends aware of things like smoke alarms and telephones to letting them know if someone is at the door.

Passing a rigorous training period, Sheena was clearly proud of her new skills and responsibility. She made a new life with a deaf woman named Hannah.

One day Hannah was out walking with Sheena when she started to cross a street. Hannah was daydreaming a bit and forgot to glance at Sheena to see if she should advance (there might be a siren in the distance that she couldn't hear but the dog could). Hannah walked toward the middle of the street, unable to hear the shouts of other pedestrians, who were yelling that a speeding car was coming toward her. Sheena tensed for just a moment, then threw herself between Hannah and the car. Hannah jerked back and escaped injury and Sheena took the blow. The car skidded to a halt leaving Sheena crumpled on the road.

Fortunately, Sheena's injuries were not too serious and she recovered after a week at the vet's. Sheena and Hannah were happily reunited once again.

He that will not be merciful to his beasts
is a beast himself.
—Thomas Fuller

Eleven-year-old Gary Watkins was playing happily and chasing lizards with Weela, his Pit Bull dog. Suddenly Weela seemed to attack Gary. She fell against him, hitting him hard. Gary's mom, who was watching, had always trusted Weela and was not sure what had happened.

Then a rattlesnake struck Weela on the face. Weela had knocked Gary out of the way of its poisonous venom. Weela recovered. Later it became apparent that this adventure was just the beginning of a lifetime of heroic acts by Weela.

In January 1993, heavy rains fell in California. A dam broke and a normally three-foot-wide river became a torrent. Weela and the Watkinses worked for six hours to rescue twelve dogs from a flooded ranch. During their hard battle against strong currents and floating debris, the Watkinses saw that Weela could sense quicksand, work through hazardous bogs, and herd others away from dangerous situations.

During the storms, seventeen dogs, puppies, and a cat were stranded on an island. There was no way to get food to them. Someone thought to put a harnessed backpack on Weela and have her swim across the river to the starving animals with dog food on her back. Although Weela weighed only sixty-five pounds, she successfully carried from thirty to fifty to pounds of food on each hazardous trip. She later aided in the rescue of thirteen stranded horses.

Weela saved people, too. Thirty people were attempting to cross a river to safety. The river had once been easily navigable on foot but had become deep, swift, and dangerous. Weela guided them to a point where they could cross without harm.

Weela, to whom so many beings owed their lives, had once been discarded. When she was just four weeks old, Weela was dumped in an alley and left for dead. She was found with her littermates and rescued by Mrs. Watkins. There was no way Mrs. Watkins could have known that one day her son and many others would owe their lives to the brave little dog.

A dog is the only thing on this earth that loves you more than he loves himself.

—Anonymous

Old Man was a chimpanzee who had been rescued from a laboratory when he was twelve years old. He was relocated to an island at Lion Country Safaris in Florida, where he lived with three females. All four chimps had been in abusive situations in their younger days. Now, no human ever approached them for fear of reprisals.

Marc Cusano, a Lion Country employee, gradually became fascinated with the chimps. He had been told they were vicious, but he grew dissatisfied with just throwing food at them from a boat—he had to try to make contact with them.

One day Marc brought the boat in close and handed Old Man a banana. Old Man took it from his hand. Old Man became friendlier and friendlier, until one day Marc was able to safely step onto the island. After a while, Marc even got close enough to Old Man to be able to groom him, the height of chimp companionability.

While working on the island one day, Marc slipped and fell. This frightened a baby chimp, who screamed. The infant's mother interpreted Marc as a threat to her infant and charged at him, biting him. Then the other two females attacked him, too. When Marc saw Old Man rushing toward him he thought that nothing could save him now.

But Old Man was there to rescue him, not hurt him. Old Man grabbed the females and threw them off Marc. He protected Marc by aggressively keeping the females away until the wounded Marc could drag himself to safety. Marc's chimp friend had saved his life.

> *The fact that man knows right from wrong proves his intellectual superiority to other creatures; but the fact that he can do wrong proves his moral inferiority to any creature that cannot.*
>
> —*Mark Twain*

Friends

Love given when it is inconvenient is the greatest love of all. Kindnesses that are shared at high cost to oneself are the most dear.

Buffy and Tawnie were both Labrador-Shepherd mix dogs who spent most of their time in the backyard. In the evening, they were invited indoors to share the life of the Gundran family. At night the two dogs would stay in the garage. As the two dogs lived into their thirteenth year, Tawnie suffered from back pain and Buffy gradually lost her sight.

Toward the end of Buffy's life, she began to suffer small strokes, yet each time she recovered enough to live on with some enjoyment. On the last night of her life, Buffy apparently wandered out of the garage and sat by the gate of the dog-run in the pouring rain. There she suffered a massive stroke.

The next morning it was still raining. The Gundrans went into the garage to say good morning to the dogs, but both were missing. When they finally searched outside, they found both dogs lying side by side in the downpour.

The Gundrans noted that the garage door had been open so that Tawnie could have taken shelter at any time during the night. But because Buffy's stroke had left her incapacitated, her dear friend Tawnie, though herself suffering from severe arthritis, chose to spend the entire night outdoors lying next to Buffy, trying to shield her from the rain and keep her warm. Both dogs were cold and soaked.

The Gundrans gently helped Buffy inside; Tawnie followed. Tawnie had suffered greatly through the long cold night but she had refused to leave the side of her lifelong buddy. "We think Tawnie knew that it was time for her friend to prepare to leave this earth, and so spent the last, long night by her side. I have a lump in my throat just writing this, but I feel I would like to share this with others," writes Ms. Gundran.

Tawnie had to consider what was most important on the last night of her friend's life. She decided it was not her own physical comfort. Even though she herself was plagued with crippling arthritis, she would not permit her friend to die alone. Throughout the long, cold, rainy night, the comfort and warmth of the garage lay just a few feet away, but Buffy could not even drag herself to it.

Tawnie proved that when someone needs you, that is all that matters.

> *I saw deep in the eyes of the animals, the human soul look out upon me.*
>
> —*Edward Carpenter*

Gail kept two fancy fantail orange/white goldfish who seemed to like each other, chasing each other and swimming together. The behavior of the two goldfish made Gail think that they had a real affection for one another, but she could not have predicted what would transpire as they grew older.

When one of the fish weakened and could no longer swim to the surface to be fed, the stronger fish would swim under him, lifting him to the surface. When the weak fish eventually died, the survivor grew depressed, would not frolic, and even ceased eating for a time.

Concerned, Gail introduced a new companion fish into the bowl, hoping to cheer him up. At first, the lonely survivor would not accept the new fish. He seemed to need time just to mourn the loss of his longtime companion. Eventually the two made friends and he regained his *joie de vivre*, again swimming and playing happily in his tank.

> *I see shining fish struggling within tight nets, while I hear Orioles singing carefree tunes. Even trivial creatures know the difference between freedom and bondage. Sympathy and compassion should be but natural to the human heart.*
>
> —*Tu Fu*

I purposely omitted from *Peaceful Kingdom* a story about a baboon named Johnny, simply because it seemed too amazing to be real. Later, however, while doing some follow-up research for another story, I found a second reference to Johnny and a referral to an old *Life Magazine* article which included numerous pictures of Johnny doing the very things I had found too incredible to believe.

In 1955, an Australian farmer called Lindsay Schmidt was driving down a country road when he encountered a roadside truck in trouble. Its driver was struggling to control a small fire that had ignited the engine. Lindsay stopped, ran to the rescue with his own fire extinguisher, and speedily put out the fire.

The grateful truck driver offered Lindsay a unit of his cargo as a reward. To his surprise, when Lindsay looked in the back of the truck, he saw twenty pairs of frightened eyes staring back at him: a group of baboons. The truck driver owned a traveling circus and used the baboons in his act. Lindsay regarded the apes curiously. Lindsay had migrated to the Australian outback only to find peace and quiet; he had no idea why he might want a baboon.

Yet at that moment, one of the baboons shyly came forward as if to greet Lindsay. It was a fateful moment. The man shared a long look with the baboon and a remarkable partnership was formed. Lindsay accepted the baboon.

When they got back to Lindsay's home, he cooked dinner and put a bowl on the floor for the baboon, whom he decided to call Johnny. Johnny looked at the table and the plate which Lindsay had set out for himself, then skeptically eyed his own dish on the floor. With a snuffle of wounded pride, Johnny picked up his dish from the floor and placed it up on the table, where he joined

Lindsay. It was the first of many indications to Lindsay that he had not obtained a pet, but a companion who clearly considered himself of equal stature.

During the next few days, Johnny curiously followed Lindsay about the house and grounds, watching him do the chores. One day, when Lindsay was trying to figure out how to pick up three heavy buckets of chicken feed, Johnny grabbed a bucket and lumbered off to feed the chickens himself. Johnny performed this task so expertly that from then on feeding the chickens was Johnny's chore.

Part of Lindsay's farm routine involved driving his tractor out to the sheep pastures where he would throw down bales of hay at regular intervals. He had to drive out to each field, stop the tractor, clamber back to the trailer piled high with hay bales, heave one out to the pasture, and then return to the tractor to start over again. It occurred to him that he could cut the work time in half if he had a worker stationed on the hay trailer who could just toss the bales out at the proper intervals while he drove. Johnny had done so well with other chores, Lindsay decided to give him a try.

Johnny rode in the trailer and quickly learned to toss off hay bales at Lindsay's signal. The baboon worked the job as if he had been born to it and the sheep feeding went much faster and easier. Things proceeded perfectly until one day when the tractor hit a buried rock. The tractor lurched, throwing Lindsay forward over the front. He slammed to the earth directly in front of the tractor.

Stunned from the fall, he looked up to see the tractor rumbling inexorably toward him—in an instant he would be crushed to death. But then he heard the motor abruptly stop and the tractor halted, inches from his body. Astonished, the farmer pulled himself slowly to his feet and peered around the big tractor engine at the driver's seat. There sat Johnny, his furry hand on the ignition key. When Lindsay had been thrown in front of the tractor, Johnny had immediately leapt from the hay trailer to the rear of the tractor, climbed into the driver's seat as he had often seen Lindsay do, and turned off the engine.

The two became best friends and Johnny even learned to drive the tractor around the fields by himself. Photographs of Johnny calmly and competently driving the tractor appeared in *Life Magazine,* along with a witness' description of Johnny's careful steering back and forth.

Johnny's status as a farmhand was eventually recognized by the Australian government. Lindsay listed Johnny as an employee on his tax returns and, after an investigation, Johnny's status as "baboon field hand" was officially accepted.

Every evening Lindsay and Johnny would retire from their hard work to drink beer together, like two working buddies anywhere in the world.

Kindness to all God's creatures is an absolute rock-bottom necessity if peace and righteousness are to prevail.
—Sir Wilfred Grenfell

Thank goodness many humans respond with great love and compassion to their animal brethren. Daphne Sheldrick has founded an orphanage in Nairobi, Kenya, to care for homeless baby elephants whose parents have been killed by poachers.

Often the baby elephants have seen their parents slaughtered with high-powered rifles, their tusks hacked out of their skulls with axes. The young elephants are so grief-stricken that they are in danger of dying of broken hearts.

The caretakers take great pains to keep the babies warm at night because they are susceptible to pneumonia—they give the little elephants blankets and actually sleep next to them. When they are old enough, the youngsters are taught how to forage for food and water themselves; the goal of the program is to return the animals to the wild.

In the twenty years that the program has been running, Sheldrick and her staff of eight caretakers have returned twelve orphaned elephants to their natural habitat in nearby Tsava National Park—a small but significant number.

When zookeeper David Gucwa discovered that his charge, Siri the elephant, loved to draw, he befriended her. He wrote the book *To Whom It May Concern* about the experience. Whenever David entered Siri's cage, she recognized him immediately. David was kind to her and his deep voice comforted her; he was not like many of the other humans she had encountered in captivity, who treated her as if she had no feelings. Siri would respond to his presence by putting her head down low and wrapping her giant elephant ear around him, in what must be one of the most wonderful greetings in all of the animal kingdom.

A recent television special on elephants showed many acts of caring among these animals. For example, a female elephant got caught in trap and couldn't feed herself. But she was not left alone to die: her elephant friends found her, fed her, and kept her alive.

The worst prison would be a closed heart.
—Pope John Paul II

A woman in a butcher shop watched as the butcher threw a bone to a large dog. The dog immediately ran outside with his prize. Minutes later, the dog was back and the butcher threw another bone to him. The butcher explained that the dog had a canine chum who, after being hit by a car, was lame and nearly blind. Every day the dog would bring the first bone to his friend and then, when he was sure his friend was taken care of, come back to get another for himself.

All of our institutions must now turn their full attention to the great task ahead to humanize our lives and thus to humanize our society.
—James Perkins

Humankind often creates a hell on earth and then curses God for its existence. But sometimes a little bit of life manages to sneak into our self-created miseries to save us.

Of all the torments this world has known, few have been as devastating to the soul as an eighteenth-century madhouse. The dungeons were black and full of horrors. The terrifying howls of the other inmates were enough to drive a sane person mad.

Into this filth, isolation, and decay strolled what seemed to be a small, furry, self-contained god. In ages past, cats had been worshipped in Egypt as divine and many have retained this regal sense of self no matter what their present-day circumstances. Drawn to the adventures offered by a rat-infested insane asylum, Jeoffry the Cat became the sole note of sanity for poet Christopher Smart during his eight years of solitary confinement. Jeoffry eased the loneliness of the poet's days and his warm body chased the demons from the long, dark nights. I do not know how Christopher Smart arrived at his miserable confinement, but the strength he took from his little cat companion is palpable in these lines from his famous poem, *To My Cat, Jeoffry,* written while imprisoned:

> *For I will consider my cat Jeoffry.*
> *For he is the servant of the Living God,*
> *duly and daily serving Him.*
> *For he keeps the Lord's watch in the night*

against the adversary.
For he counteracts the powers of darkness
by his electrical skin and glaring eyes.
For he counteracts the Devil, who is death,
by brisking about the life.

Christopher Smart was released from the asylum in 1764. The only love he had shared, the only being he had trusted, the only kindness he had experienced during those eight years, had been with a small, street-smart, rat-catching cat he named Jeoffry.

Looked at rationally, we might doubt that having a small, somewhat helpless being of another species in our home could be indispensable. Animals do not, after all, make any money for us, converse with us so that we may broaden our intellects, or improve our sex lives. So why does over half the population maintain a companion animal?

We might as well ask why we like spring. Or moonlight on water and the unexpected scent of jasmine in the night air.

The other day, I was teasing my cats that they were completely useless because they just lay around like furniture while I worked on this book. My husband overheard me and replied, "Ah, yes, but when you are lonely, it does you no good at all to rest your head against a piece of wood."

In the company of another being, we find life and hence sweetness, a connection to an Eden lost but longed for within another's other's touch and breath. Life must have life to love, else it sours upon itself. By being an angel to an animal, we can bring out the angel in ourselves—the angel we might otherwise never have seen. When we offer the gift of love to another, we find more within ourselves. Love is both ephemeral and real. It is spirit, yet apparent. It cannot be seen, but its lack can wreck a life.

Grace Under Fire

Animals speak to us most powerfully through actions so astoundingly courageous that they practically qualify for sainthood. Here is the story of a dog who would neither give up nor simply save his own skin (or fur, in this case). Instead, he risked everything when the person he loved was in danger.

The temperature had fallen almost to zero one night in December 1964. It was so cold and icy that Marvin Scott worried there might be damage to a small boat moored to a floating dock near his home in Washington state. Marvin decided to check on it.

Patches, his Malamute-Collie dog, followed Marvin as he trudged down to the dock. Sure enough, ice was forming on the boat. Trying to fix the situation, Marvin braced himself on the dock and pushed on the stern line with a stick. But the boards of the pier had become slick with ice and his legs shot out from under him. He fell from the pier and slammed full force against a floating dock, tearing almost all of the muscles and tendons of both legs.

In shock and agony, Marvin hit the frigid water and went under.

Thrashing desperately in the stunningly cold water with the weight of his own heavy clothes pulling him down, Marvin knew his chances of survival were extremely slim. Desperately he struggled to reach the surface, but continued to sink downward into the numbing depths.

Then something grasped him by the hair and pulled him up to the surface. Marvin gulped for air and saw that it was Patches, who had jumped into the icy lake and was now vigorously pulling him by the hair. Patches swam with all his might, towing the injured man towards the floating dock.

When they reached the dock, Marvin helped push Patches onto it. But when Marvin tried to climb up, he discovered that his legs had been too torn up by the fall and no longer worked properly. Gathering all his strength and will, Marvin made one last attempt to heave himself up, but the effort was too much and he blacked out.

Patches again abandoned the safety of the dock and dove into the water after Marvin. Again the brave dog pulled his friend up by the hair until he could, once again, grasp the dock. Marvin held onto the side of the dock and screamed for help but although his house was only 300 feet away, the wind overwhelmed his cries. But Patches had not given up yet and Marvin felt the dog's teeth pull at his overcoat as he tried to drag Marvin onto the dock. Combining their strength, Marvin and Patches both managed to climb onto the dock. Almost overcome with pain, Marvin tried to crawl. With Patches dragging him as well, the two of them traversed the 300 feet to the house.

Marvin's recovery took many months. He knew he owed his life to the amazing dog who refused to consider his own safety or suffering in order to help a friend. Patches was named the Ken-L-Ration (now Reward) dog hero of the year.

TO A DOG

If there is no God for thee,
Then there is no God for me.
—Anna Hempstead Branch

Animals who are not even a part of one's own family will often risk their lives to protect someone who is clearly helpless in a dangerous situation. One little boy was very lucky that a friend of his family brought along a dog companion on an outing.

At first it was just a pleasant day in the country for young Philip Stevens. The toddler's mother was close by as he played in the grass with James, an older boy, and James' dog, Rusty. Perhaps because it seemed so pleasant, no one was watching him with too much concern. When Philip went off to play alone in a barren spot between some hedges, he appeared quite safe.

Then something frightened a group of cows grazing nearby. They turned and stampeded toward the only way out of the field, a break in the hedges—the very spot where Philip was playing. His terrified mother tried to run to him but she saw that she would never make it in time. She screamed at the realization that her son was about to be trampled to death.

Rusty also saw the danger and raced toward the toddler, barking insistently.

The cows galloped wildly toward the gap—but now a determined dog stood in their way. Rusty had run to the break in the hedges and stood her ground directly in front of Philip. The running cows had to veer to each side to avoid the loudly barking dog. A stream of cows raced past Rusty and Philip on either side, the wind from their galloping bodies riffling the boy's hair. Rusty valiantly protected the boy, standing her ground even as she was kicked in the shoulder by one of the stampeding cows.

When it was all over, Philip was shaken but fine, and the courageous Rusty was awarded a special medal for bravery.

> *Be careful that the love of gain draw us not into any business which may weaken our love of our Heavenly Father or bring unnecessary trouble to any of His creatures.*
>
> —*John Woolman*

Sixteen-year-old soldier John Granville was killed in battle and buried in a common grave. Sixteen is an awfully young age to die.

Apparently his faithful dog thought so, too; the dog dug young John up again. John, as it turns out, was still alive—if only just barely. He recovered and lived another fifty-seven years.

In a letter to *People Magazine,* Mary Anne Surran wrote that animals had saved her two children. The first child was saved from choking when the family's cat meowed and drew attention to her. The second child was saved after he fell into an anthill. Their dog pulled him from the anthill and then tugged the child into a small pool to get the ants off him.

Sometimes love can come from where one least expects it. And as this story shows, so can desperately needed assistance.

It is every woman's nightmare to be attacked while alone at home. Barbara screamed for help when a drunken stranger forced his way into her home in Thomaston, Connecticut. To her terror, she saw there was no one around to hear her screams—no human, that is.

Suddenly Barbara's African gray parrot, Samantha, came screeching into the room. As Barbara screamed, Samantha flew at the man's face and struck him with her beak and sharp claws.

The man stepped back in confusion, struggling to fight the tiny defender off, only to howl in pain as the fangs of Gomer, Barbara's Chihuahua, sank into his leg.

Fighting against the two tiny warriors, the drunken stranger yelled and flailed about ineffectually, finally fleeing back into the street.

The parrot Samantha weighed under a pound, and Gomer was only a small Chihuahua, but sometimes bravery and bravado win the day. Barbara's rescuers proved that big hearts and great commitment can come in very tiny packages.

> *Ever occur to you why some of us can be this much concerned with animals' suffering? Because government is not. Why not? Animals don't vote.*
>
> *—Paul Harvey*

In Lentino, Italy, a three-year-old girl named Palma wandered into the street. Bystanders froze in momentary horror as they saw a truck speeding toward the child. A quick-thinking parrot hurled itself at the windshield of the truck, causing the driver to hit the brakes, and thereby saved the life of the child.

In another case, a band of thieves made the mistake of talking to each other during a heist, even calling each other by name. In the shop, unseen by them, a parrot quietly listened. When the police arrived, the parrot cooperatively recited parts of the gang's conversation and their names. The thieves were quickly captured.

> *I'll not hurt thee, says Uncle Toby, rising with the fly in his hand. Go, he says, opening the window to let it escape. Why should I hurt thee? This world is surely wide enough to hold both thee and me.*
>
> —*Lawrence Sterne*

Robins in the wild will fight each other in fierce battles over territory. But the victorious robin does not simply abandon the victim to his fate. In fact, the victor will often stay and feed his conquest, even if it means not migrating in the winter.

Sometimes love really does conquer all.

Mooch was the laughing-stock of his firehouse. Supposedly, the dog was the mascot of Fire Engine Company No. 11, but he was terrified of fires and would run in the opposite direction if he smelled smoke.

But there are circumstances in everyone's life that lead one to put aside fear for one's own safety and follow the path of courage. For Mooch, such a moment came when his sweetheart, a female dog named Winkie, was in danger.

One day a building close to Mooch's firehouse caught fire. Winkie was trapped inside. Her desperate yelps aroused Mooch into a frenzy of concern, driving him to overcome his own fears. Mooch ran into the flaming building, ignoring his fear of smoke and the deadly flames. He found Winkie, dazed with heat and exhaustion, and dragged her outside to safety.

Winkie recovered and Mooch, a true and valiant hero, received a medal for bravery.

Cleveland Amory tells the story of a ten-month-old kitten who eventually came to be known as Matt.

A hiking party was preparing to climb the Matterhorn. One climber in the party had a kitten. The man arose early to keep his appointment with the massive peak, leaving the kitten behind in his hotel room. He and his party dressed in their hiking gear and began their ascent. They carried food, water, rope, mountaineering equipment, and emergency supplies.

After much strenuous effort, the group finally made it to the summit. They cheered and congratulated each other, joyous that they had overcome many difficulties and persevered when it seemed they could not take another step. But the feeling of triumph, of humanity conquering nature, of going where (almost) no one had gone before, was a bit dampened when they heard a pitiful mewing sound.

Apparently Matt the kitten had become very lonely and a little confused as to why his human friend had left him at the hotel all alone. Perhaps something was the matter and he should just go see. So Matt finally decided to try to find his friend and followed after him up the trail. The kitten had climbed to the top of the 14,780-foot peak—*without* food, water, or equipment.

Profound courage never hesitates. Here is a story of a dog's split-second heroism that occurred many years ago in New York City.

The fire truck raced through the city streets, sirens wailing. Driving toward the blaze at top speed, the fireman suddenly noticed a three-year-old boy frozen in the middle of the street, too terrified to move. The driver slammed the brakes and the huge truck gradually slowed with a great screaming of brakes, but it had too much momentum to stop quickly enough to avoid hitting the boy. The driver watched in helpless horror as his truck plowed forward, screeching, about to roll over the boy within the next three seconds.

Then one of those incredible canine miracles happened.

Jack, the Dalmatian mascot of Brooklyn New York Engine Company 105, was sitting in his usual place next to the petrified driver. Instantly, the dog's keen eyes assessed the situation and he decided to give his all to reach the boy before the truck. Jack leapt out of the still-moving vehicle, raced ahead, and with all of his determination, leapt into the path of the onrushing truck, knocking the child clear. The dog and boy rolled to the side by the curb as the truck swept past and finally came to a full stop. Jack and the boy both missed being crushed by inches.

Jack received the Medal of Valor from the New York Humane Society.

For all his learning or sophistication, man still instinctively reaches toward that force beyond. Only arrogance can deny its existence, and the denial falters in the face of evidence on every hand, in every tuft of grass, in every bird, in every opening bud, there it is.

—*Hal Borland*

Cats often get a bad rap where acts of heroism are concerned, so it may surprise you to find stories of cat heroes that reveal a remarkable type of feline bravery that is more common than you might have suspected. According to the Reuters news service, a thirty-three-year-old woman in San Diego was recently attacked by a man who gained entrance to her home by picking the lock to her door while she slept.

Suddenly some demon-from-hell—the stranger knew not what—pounced on his shoulder. The man shrieked as sharp claws sank into his arms and back, furiously ripping his flesh. He yelled, and as the attack continued, the woman awakened. The burglar put his hand over the woman's mouth and ordered her not to scream. She bit him and he fled.

Jake, her eighteen-year-old cat—an age at which a cat should be completely retired—had acted quickly, attacked the intruder, and saved the woman. There are many similar accounts of cats saving people from burglars.

> *If only I could be presented to the emperor, I would pray him, for the love of God, and of me, to issue an edict prohibiting anyone from capturing or imprisoning my sisters the larks, and ordering that all who have oxen or asses should at Christmas feed them particularly well.*
>
> —*St. Francis of Assisi*

You might assume that carrier pigeons fly home strictly because of instinct and that character and determination really have nothing to do with it; that this behavior is just the bird's genetic programming. In fact, pigeon fanciers know it is extremely important that the carrier pigeon have a strong motivation to return home and that the bird's will is powerful.

When released, some birds will fly into the nearest tree as if to say, This is good enough, thank you very much. I will make my new home here. Nothing will persuade them to move on, no matter how important the message they are carrying. In other cases, the birds seem to know that it is imperative they make it back, and do so despite broken wings, gunshot wounds, and attacks by hawks.

One pigeon, badly wounded in flight, *walked* several miles back to camp. Sometimes it is love that motivates the pigeons to fly through battlefields and bombings. The pigeon keepers know that certain birds really love their mates and will literally fly through wartime hell to get back to their true loves in record time. Like human athletes, it seems the difference between the good and the great lies in putting all of one's heart into the goal.

In October 1943, the U.S. Army's 169th Infantry Brigade captured the town of Covi Vecchia, Italy, which had been a Nazi stronghold. Unfortunately, Allied pilots, unaware that their fellow Americans had taken the town, were preparing to launch a bombing attack on the city.

The fate of the brigade rested on the wings of GI Joe, a carrier pigeon who was flying toward headquarters with the news of the Allied capture of Covi Vecchia. Even as the planes and pilots were readied for take-off, GI Joe was flying across the twenty miles between the town and U.S. Army headquarters.

Twenty miles is a long journey at top speed, even for a brave little bird, but GI Joe flew as fast as he could and made it in time—the 169th Brigade was saved. Had he been even five minutes longer, the planes would have taken off and there would have been no way to call them back.

GI Joe later received the Dicken Medal at an investiture held at the Tower of London in October 1946.

Nicole Russell was backing out of her driveway when four strangers approached her car.

"Give me your key!" one shouted roughly.

"Just take everything," she cried. She hoped she would just be robbed and not harmed. But then the four men started pulling her toward their car.

Nicole started screaming for her life.

Her family rushed outside to see what was wrong. And one important part of her family was a small Staffordshire Bull Terrier named Bella.

The rest of the family froze in momentary panic as to what to do. Then one of the men tried to stab Nicole. Bella rushed at him, grabbed him by the leg, and would not let go.

Another man drew a gun and fired: a bullet pierced the little dog. Severely wounded and in pain, Bella refused to give up. She fought the men with everything she had. She knew that Nicole was completely depending on her and she would not be stopped.

Despite the bullet wound Bella attacked again. This time she was victorious and the men finally fled.

Nicole was safe, but Bella died from her wounds. She gave her life for the person she had loved better than life itself. At a sad ceremony held in remembrance for the courageous little dog, Bella received a posthumous trophy for her brave actions.

> *I think that dogs are the most amazing creatures; they give unconditional love. For me they are the role model for being alive.*
>
> —*Gilda Radner*

The Things We Do for Love

"Fighting like cats and dogs," the saying goes. And yet friendship is a mysterious thing and love can conquer separations—even those between species. And if even separate species can learn to love, why not us humans?

A cocker spaniel named Duke was going blind. He had gotten on in years and his formerly sharp senses were failing him. Mrs. Roundtree, with whom Duke lived, looked out her window one day and saw the dog wandering outside, confused and unable to find his way back to the house. Worried, she was about to go out to help him when she saw her cat, Bos'n, race toward the disconcerted dog. As Mrs. Roundtree watched, Bos'n took control. The cat nudged the dog first on one side, then the other, gently guiding him back to the house.

From that day forward, Mrs. Roundtree never had to worry about Duke again. Bos'n took assiduous care of the blind dog and was always at the ready to guide him around. The dutiful cat not only helped Duke when he seemed about to wander off, but also kept a sharp eye out for potential trouble. He protected Duke, helped him move from one place to the other, and would never let Duke get too close to the edge of the porch, where there was no railing, for fear that he might fall. A cat was truly a dog's best friend.

In England there is a retirement home for unwanted animals where nearly a thousand animals have come to live. The owner of the home, Kay Lockwood, says that the blind animals are always looked after by the others. When a blind donkey wants to go for a walk, there are always two with good eyesight who position themselves on either side of the blind donkey so as to guide him safely.

Charles Darwin learned of several crows who fed one of their own kind after the bird had lost its eyesight.

> *It does not necessarily follow that a scholar in the humanities is also a humanist but it should. For what does it avail a man to be the greatest expert on John Donne if he cannot hear the bell tolling?*
>
> —*Milton Eisenhower*

Sometimes an animal's concern is completely unwarranted, yet their dedication to their task is so charming it is still noteworthy.

Sherry Kuring of Cucamonga, California, discovered she did not have to worry about leaving the baby alone briefly. If anything disturbed the infant, Sherry's Chow-Retriever mix Drifter would run to the next room, lunge against the door, and utter one emphatic *woof!* He then would hurry back to the nursery, confident that the mother was following him.

Sherry's cat Bailey was also a stickler for good childcare. When Sherry was changing her baby's diaper and the baby began to cry, the alarmed cat—who must have figured Sherry was harming the child—leapt upon Sherry's bare leg, wrapping both front paws around it, and gently pressed her teeth against the flesh as if warning, "Don't you even *think* about hurting that baby!" Bailey continued to repeat this warning whenever the baby cried at diaper changes or feeding, eventually attenuating the warning bite to a gentler, you-better-not-hurt-our-baby lick.

Lessa, an Australian Shepherd dog, would not come when called. Instead, he made a fussing noise in the backyard. Sue Corbett went to see what the trouble was. Lessa was occupied with a baby possum that had gotten separated from his mother. The dog kept nudging the small white form. Sue snatched the baby possum from the dog and put it near a warming oven in the house to keep it from catching a chill. But Lessa proved genuinely concerned for the small refugee. Whenever Sue left it by the warming oven, Lessa would stay and guard it, lovingly licking the helpless creature.

Later Sue put a burrito for her mother inside the same warming oven, next to the young possum. The mother came, opened the warming oven, and bit into the narrow white burrito. Seeing this, Lessa became hysterical, barking wildly, and would not calm down. "What have you done with my baby possum?" she seemed to be saying. The dog fussed and cried out until she was shown both the burrito *and* the baby possum and realized that all was well.

> *Pretending to be the king of nature, man has only brought it to poverty. The time has come for man to become an elder brother in the family of nature.*
>
> *—Dimitrii Sukharev*

Dr. W. F. Sturgill, a physician for the Norfolk and Western Railroad, once treated a dog for wounds suffered from entanglement in barbed wire. The dog recuperated and was released.

A year later, Dr. Sturgill heard a scratching at his door and went to investigate. The dog he had treated the year before was outside his door, this time without his human companion. But he was not alone. He had brought along with him a strange dog whose paws were bruised and bleeding. Both dogs looked to the good doctor expectantly, as if asking for treatment.

Somehow the dog had remembered that this was the man who could fix hurt paws. Dr. Sturgill treated the new dog's hurt paws and the two dogs walked off together.

> *Kinship with all creatures of the earth, sky, and water was a real and active principle. For the animal and bird world there existed a brotherly feeling that kept the Lakota safe among them, and so close did some of the Lakotas come to their feathered and furred friends that in true brotherhood they spoke a common tongue.*
>
> *—Chief Luther Standing Bear*

On a spring day, Don and Layzette sat eating lunch by a beautiful, partially iced-over lake. They noticed a lone duck swimming in an open area and wondered why it had been left behind when the other ducks had migrated.

A pair of bald eagles flew overhead and began diving at the duck. The alarmed duck dove beneath the surface repeatedly to avoid their attacks. When the duck did not fly away from his attackers, the couple realized he could not fly and that he must be injured—no doubt this was the reason he had been left behind.

Nearby, two swans noticed the duck's plight and swam toward him. Then, to the astonishment of the two human observers, the two swans began shielding the duck from the eagles' attacks. Whenever the eagles dove at the duck, the swans flapped their wings and skittered across the water, driving the eagles away.

The swans herded the duck away from the ice and toward a wooded point of land. Although the duck could not fly at all, under the swans' urging, he managed to climb out of the water and into the safety of the woods.

We must meet the forces of hate with the power of love:
we must meet physical force with soul force.
—Dr. Martin Luther King, Jr.

An eagle was shot and lay wounded upon the ground. He was rescued by a man who loved animals. The bird's wing was hopelessly shattered; he could never fly again. The man, who loved hang-gliding, well understood the anguish in the heart of the crippled eagle.

There was only one thing to be done. The man fitted a special harness on his hang-glider. Now the two of them go soaring together, and both enjoy the eagle's-eye view of the world.

One lesson, Shepherd, let us to divide,
Taught both by what she shows and what conceals;
Never to blend our pleasure or our pride
With sorrow of the meanest thing that feels.

—*William Wordsworth*

Grey Owl was the name an Englishman took after he immigrated to America, lived amongst the Indians, married an Indian girl named Anahareo, and made her people's ways his own.

Grey Owl was a trapper. One day he found two tiny beaver kits and took them home to raise them. The baby beavers were so full of life and mischief and fun that he realized he could no longer cause his fellow creatures to suffer painful deaths. Grey Owl gave up trapping beaver forever.

McGinty and McGinnis, as he named the two kits, grew up to be large and happy beavers, well satisfied with their new home. One day Grey Owl brought home a large pile of moss which he placed on the floor, expecting to stuff it into the cracks between the logs of his cabin for insulation. That night Grey Owl and Anahareo went to sleep. But beavers are great builders of dams and they know just what to do with moss and logs. When Grey Owl awoke the next morning, the spaces between the cabin's logs had been expertly stuffed as high as the beavers could reach.

Not all beaver talents are useful to people, however. Grey Owl had a tall table upon which he and Anahareo liked to keep their food—food that beavers might like to eat if only they could get to it. One day Grey Owl and Anahareo returned home to find that all four legs of their table had been gnawed down and now the table was at a nice lower height—just right for beavers to dine.

To help life reach full development the good [person] is a friend of all living things.

—Albert Schweitzer

When baby monkeys lose their mothers, they are frequently adopted by other members of their tribe. The males as well as the females seem to take a special interest in the little orphans and see that they are protected and cared for. While human beings tend to think that their finer emotions are not shared by other animals, Professor J. Howard Moore feels differently: "In the gentle bosoms of these wild woodland mothers glow the antecedents of the same impulses that cast that blessed radiance over the lost paradise of our own sweet childhood."

In *The Universal Kinship,* Professor Moore describes the love of a monkey mother as she leaps from limb to limb, stuffing green leaves into her child's wounds, trying to stop the flow of blood from the hunter's bullet lodged there. Fleeing the hated hunters, she embraces her child, until another bullet sends her plummeting to the ground. She had in her simian soul mother-love as genuine, and as sacred, as that which burns in the breast of woman.

> *I've always felt you don't have to be completely detached, emotionally uninvolved to make precise observations. There's nothing wrong with feeling great empathy for your subjects.*
>
> *—Jane Goodall*

If we all are truly connected, if we all are strands in the web of life, then why should we ever be surprised to find evidence of love in other hearts? Why should not every feeling known to man or woman, whether in rudimentary form or complete, be found in the animal kingdom as well? And if animals are incomplete while we are whole in our emotions, why not let ourselves feel toward them the natural tenderness and protection that one would have for an immature brother or sister?

Why does humanity seem to feel such a need to separate itself from the animals, to the extreme degree that some, especially in scientific circles, do? We espouse Darwin's theory of evolution and, at the same time, act as if we are ashamed of our relatives. "Just keep old Aunt Harriet locked in the attic," we seem to be saying. "She might embarrass us."

Yet it is the angelic nature of all beings, the part that is good and loving and even self-sacrificing, that compels our notice. Throughout the animal kingdom we come across the signature of the Creator who bestowed loving kindness in the heart of every creature.

A Lifetime of Love

Tender-hearted people are often teased if they can't refuse to take in stray animals. Over time, adopting strays can become too much of a good thing.

But what are we to do when yet another starving dog or cat comes to the door? Why is it that we, as a society, think we have to tease those we consider overly kind? Isn't it be better to be over-kind than under-kind? This next story could be titled: Revenge of the Little Old Lady Whom Everyone Thought Had Too Many Cats.

Mrs. Nina Sweeney was an elderly lady who sometimes took in cats and other animals. Many a stray had been fed by her, many a starving creature, for whom she was their last hope, had been met with kindness and a bit of food and love when they came begging at her door. She often found homes for the stray animals and was finally able to reduce her rescued animal population to seven cats and one dog. The animals loved her for her charity and one cold night they were able to repay her for all that she had done.

In January 1953, Mrs. Sweeney fell ill and was unable to move from her bed. The fire she lit in the stove went out after awhile and the house grew terribly cold. She was very feeble and realized she would probably freeze to death before help came. She knew that, in the past, this had been the fate of a number of old women living alone.

The next day, a neighbor arrived to check on her. The house was bitterly cold and the neighbor raced to the elderly woman's bedroom, fearing the worst.

Mrs. Sweeney was safe in bed, very much alive—and not cold at all. In fact, she was quite comfortable, with seven cats and a dog draped over her. Her rescued animals had saved her life by keeping her warm throughout the night.

A growing body of research shows that companion animals can actually accelerate the healing process, improve your outlook, and even extend your life! As a result of these studies, the legal system is finally beginning to acknowledge the legal right of senior citizens and recuperating patients to keep animals as companions, even in residential complexes that forbid them.

An old woman's only family after her husband's death was her devoted cat of twelve years. She had anticipated that no matter how lonely life became, she would always be able to count on her sweet kitty making her feel better, just by curling up in her lap and purring. A true friend to the end, her cat gave her all his love and devotion, yet his own requirements were small.

But when her apartment building was sold to a corporation with a "no pets" policy, the woman was forced to either move or have her beloved cat destroyed. She searched and searched but could not find another apartment she could afford. On the day she was eventually forced to have her cat put to sleep, the last flame of love in her life was extinguished. The human cost of a "no pets" policy is immense. Tragically, stories like this are repeated thousands of times every year throughout the country.

Humanity advances only as it becomes more humane.
The highest known form of friendship is that of a dog to

his master. You are in luck if you can find one man or one woman on earth who has that kind of affection for you and fidelity to you.

—Dr. Frank Crane

Many people who live alone, especially disabled or senior citizens, have found the critical solace and spirit to go on, thanks to the love and companionship of their animal friends. Companion animals not only measurably reduce stress and anxiety, they also provide the pure unconditional love that is so necessary and yet often in short supply from human sources.

Doris Day has led the fight for the rights of seniors to keep companion animals for the sake of their psychological well-being. Ms. Day knows from personal experience how much the tender love of a nurturing animal angel can mean to a shut-in. Describing her convalescence after a horrible automobile accident, she wrote: "My constant companion was my doggy 'Tiny' and he taught me how much love, affection, and undemanding companionship a dog can give."

Doris Day's experience is widespread. In a study at the University of Minnesota, Dr. Michael Robin found that 97 percent of teens raised with companion animals considered it a best friend or someone to love. In fact, for lonely children as well as for seniors and handicapped people, an animal may be a person's greatest single source of comfort, companionship, and love. The therapeutic value of an animal's unconditional and unflagging love can be critical.

While the psychological value of a companion animal has long been recognized, only recently have scientific studies demonstrated that animals offer physiological benefits as well. A study conducted by Erika Friedmann, a biologist at University of Pennsylvania, found that heart patients with companion animals showed a higher rate of survival. At the same university, associate professor of psychiatry Dr. Aaron Katcher found that people suffering from high blood pressure live longer and better with animal friends. At a recent American Heart Association annual meeting, it was reported that blood pressure and heart rate were reduced by simply petting a dog.

In 1991 in Savannah, Georgia, seventy-five-year-old widow Gertrude Cozard found two stray and starving kittens outside her door. They were so sweet and helpless she couldn't help giving them a bit of food, as well as her wishes that they might get a good start in life.

The directors of her condominium complex sued her for $750 for feeding the two kittens.

Although she did not adopt the strays and quickly found homes for them, ending the cause of the dispute, the board nevertheless proceeded to sue her for the cost of their legal consultations.

When they understand the facts, many landlords can find room to compromise on the question of companion animals. When Doris Day became a partner in the Cypress Inn in Carmel, she immediately changed the policy to: Pets Welcome. She reports that hundreds of quadrupeds have visited the inn without a single problem.

Some wonderful changes are being made in the courts. The recent case of Durand Evan of Sacramento, California, has set a precedent for a disabled person's right to keep a "therapeutic cat." Mr. Evan, who suffers from fibromyalgia, lived alone with his cat Bammers, whom he was loathe to give up. When he moved to the low-income River Gardens Apartments, he ran afoul of their no pets policy and was threatened with eviction. The U.S. Department of Housing and Urban Development charged the landlord with discriminatory housing practices.

Mr. Evan won his case! The landlord was ordered to pay damages of $5,758 to Evan and $5,000 in civil penalties to HUD (HUD v. Dutra, DEC 226.40). Bammers stayed on.

Animals often intuit when humans most need comforting. Barbara and Stuart were robbed one evening while they were attending the ballet. As often happens, the pain of losing some beloved belongings was only half the loss. The other half was in the feeling of violation and the loss of trust in the human race. Barbara sat down on her bed feeling totally demolished, and burst into tears.

Mr. Mo, Stuart's cat, jumped up on the bed with her and cradled her face in his paws—talons withdrawn. Barbara wrote me, "Mr. Mo let me know something was still right with the world. I will never forget that moment—it comforts me still."

Mr. Mo has since passed on, but his legacy of love continues in the renewal of faith and comfort he gave.

Foxy was a small but mischievous little dog. She really liked ringing the old bell outside the home of eighty-four-year-old Mrs. Robinette of Virginia. Scolding didn't seem to help, so finally Mrs. Robinette tied a knot in the bell's rope, which suspended the rope-end three feet from the ground, too high for the playful little terrier to reach. Sure enough, that stopped Foxy from ringing the bell.

One day Mrs. Robinette took a bad fall. Too weak to call very loudly for help, she lay in pain where she'd fallen. Foxy rushed to her side, eager to help—but how? They both looked at the bell, which sometimes had annoyed the neighbors enough to attract attention. But it was hopeless now. Mrs. Robinette knew she had made the rope too high for the terrier to reach.

But Foxy gathered all of her strength and leaped into the air. She caught the end of the rope in her teeth and rang for dear life, creating a real commotion. The neighbors came and Mrs. Robinette's life was saved.

Here is a fascinating first-person account of one man's ordeal as he went into insulin shock and escaped death, thanks to the brave efforts of his beautiful Valiant Collie, Silver, who proved to be a true animal angel. This happened on March 30, 1993.

"Why is Silver pawing me? It's too much effort to think and I start to recede back into the comfortable dark. A nose rubs under my neck and I mumble incoherently. I just wish she would go. Determined claws paw at me and Silver barks. Silver rarely barks. I should concentrate. Something's wrong. I've got to think. I can't think. Silver claws me again and something clicks in my mind. I'm low on sugar. I've got to get up. If I just lay here I'll die!

"I try to push myself off the bed but my arms don't want to respond properly. I get half off and fall over sideways. If only I could concentrate I could do this. Silver is standing next to me and I will my uncoordinated arm to go around her and manage to heave myself from the bed. My legs won't hold me up—they buckle and collapse under me, pitching me towards the floor. I have to concentrate. I've got to make it upstairs to the kitchen and get something in me. I can't stop now!

"Silver routs me with her nose, pushing me, shoving me, urging me on. I crawl with the wild, uncontrolled swimming motions of a person suffering from muscular dystrophy. Its hard to walk upright, but if I can just focus.... There are the stairs ahead of me and Silver is right beside me. I can't really see her. It takes all of my concentration to see the stairs ahead, but I can feel her there.

"The stairs, at last. How am I going to climb them? 'One step at a time,' I think, as humor sidles around my sugar-deprived brain. I wrap my fingers in back of one plank step and lunge forward, falling sharply against the nearest wall. Oh, God! Don't let me bounce, if I fall over the other way I won't be able to get up again. No! I'm going the other way. I can't control my body. Silver's solid shoulder meets mine and I stay upright and come more gently back against the wall.

"The world's fading and I've got to concentrate. If I pass out now without sugar I'll die. Another step. The same routine: bounce off the wall, and then Silver. Another and another. Silver is climbing the stairs slowly beside me. We're almost to the top.

Phone. I've got to make it to the phone. Dial 911. That way if I don't get enough sugar before I pass out they'll find me.

"Last step. No wall to bounce off, only Silver. If only I can get my arm around her shoulder. Yes! Forward, Silver, forward. I can't do this much more. Yes! We're up! No! Don't pass out! A few more feet, I knock over a chair. If only I could control my muscles. The phone. I grab it. Four tries and I've reached 911. 'Send ambulance,' I try to say, but I'm too far gone to be able to speak clearly and I'm about to throw up."

The medics responded and the man's life was saved, thanks to Silver.

On September 15, 1998, Clem Brase was walking from his bathroom to his bedroom when he suffered a stroke. As he fell, Clem dropped a lit cigarette. When he found himself unable to get up, his cat, Buffy, jumped onto the nightstand and knocked the ashtray down with a determination that made Clem think it was done just so that Clem could safely put his cigarette out.

Buffy stayed by Clem's side and rubbed against his arm and hand to remind him he was not alone. She also helped keep the man conscious and awake.

The telephone rang. Clem knew whoever was calling could help him, but he didn't have enough control over his body to rise and pick up the phone. Again Buffy jumped onto the nightstand, this time knocking the phone onto the floor. Then the cat pushed it over to Clem so he could speak to the person on the other end. It was Clem's daughter; she called a neighbor who rushed over to help him. When the neighbor came to the door, Buffy rushed to greet her and escorted her to Clem.

Buffy won the Whiskas Vitalife Award for most heroic cat of the year. Whiskas has a videotaped testimonial from Clem about his heroic animal on file. "Without little Buffy," said Clem in his understated way, "I would have been who knows where?"

A man named Claussen was fearful because he was going blind. He was older and had been a cowboy all his life. He had no idea how he could round up cattle without his eyesight, and there was nothing else he had been trained to do.

However, Claussen's faithful horse Shotgun seemed to be able to guide the man and helped him to continue to do what he had always done. Riding Shotgun while blind, Claussen learned to become even more attuned to the horse than before. He found that there were subtle differences in the way Shotgun reacted to a steer, a man, or a dog. Man and horse learned to read each other well. Though Claussen is blind, thanks to his seeing-eye horse he still rides the range. Shotgun keeps an eye out for cows that Claussen cannot see as the two trot off to round up strays.

> *What is the use of this fuss about morality when the issue involves only a horse? The first and most difficult teaching of civilization concerns man's behavior to his inferiors. Make humanity gentle or reasonable toward animals, and strife or injustice between human beings would speedily terminate.*
>
> —*Dr. Edward Mayhew*

Seventeen-year-old Jeremy was dying of cancer. He knew that he would die soon and told his devastated mother, Lana, his last wish. He wanted her to have his German Shepherd, Grizzly, who had helped him through the worst times and the worst suffering. "Maybe there'll be some way you and he can help other kids," Jeremy said.

When her son died, Lana felt her whole world collapse. She became too depressed to even go out of the house. But Grizzly knew she had to go on, so one day he brought her one of her running shoes and then rifled through the closet and brought her its mate. When she protested that she just couldn't move yet, he pulled at her sleeve. She started going out into the fresh air again.

Lana began taking Grizzly to the pediatric floor of the Health Center where she did volunteer work. Perhaps because of the time he had spent caring for Jeremy, Grizzly seemed to know just how to be gentle with sick children. The children loved him. For years he brightened their days and took their minds off the physical and emotional pain they were suffering. Even kids who fought with other children calmed down around Grizzly.

One day it was discovered that Grizzly himself was handicapped. No one had known it, but he had been blind for years. When Lana asked how he could do such amazing healing work while blind, a friend replied, "He sees with his heart, not his eyes."

So many gods, so many creeds,
So many paths that wind and wind,
While just the art of being kind
Is all the sad world needs.

—Ella Wheeler Wilcox

Nurturing Nature

Animal instinct has been used to explain away many actions that, had they been done by humans, would have been recognized as acts of caring and compassion. When we humans hear of a person who has adopted an abandoned orphan, especially one with some problems or deficiencies, we acknowledge the person's kindness. Yet when an animal adopts the wounded of another species, we attribute the act to the animal's confusion.

There are many stories in which the behavior of the adopting animal clearly indicates that she knew the youngsters were not of her own species, yet cared for them anyway. A naturalist who gave a chicken several guinea fowl eggs to mother was amazed when she hatched and raised them and then led them to food which was appropriate for baby guinea fowl but not for baby chicks. Another naturalist gave a hen some duck eggs to look after. When the ducklings were old enough, their mother took them to the water and encouraged them to swim. What other explanation can there be for this behavior except that the hens knew perfectly well that they were not dealing with babies of their own kind and rather brilliantly made allowances for the differences between the species?

A dog named Sidney was due to be put to sleep at an SPCA. Then a box of four hedgehog babies were brought in. Sidney immediately adopted them, cuddling them and giving them protection and warmth until they were old enough to be on their own. News of Sidney's kindly nature reached the press and soon he himself was adopted.

Jackie, a dog who lived during wartime, was friends with a cat. He carried her kittens to the shelter for her every time there was an air raid. Kalli, a dog who lived in a wildlife park, raised baby lion cubs, foxes, pumas, goats, a badger, and a leopard. And one extraordinarily tolerant cat named Sophie fed and cared for two orphaned squirrels along with her own single kitten. The squirrel babies took to clinging to her stomach as she walked, which was mighty uncomfortable for the cat. Nevertheless, she put up with it good-naturedly—for the little squirrels' sake.

Dogs have been known to adopt and feed baby kittens. Cats have been known to raise helpless baby mice. Apparently the "law of the jungle" and natural animosity goes only so far in the animal kingdom. When there is need in a helpless young one, animals are capable of stretching their hearts to fill it.

I find this particularly stirring because we all know that dogs chase cats and cats eat mice, while hedgehogs and squirrels usually just ignore each other. This is incontrovertible. And yet, a great human story is made memorable when the people involved act with uncommon valor. If animals know what they are doing when they adopt others who are not of their own kind, simply because of the younger animal's desperate need, then they are acting *beyond* instinct. Not just in spite of it, and certainly not because of it, but above and beyond it.

That animals can act from a higher and more compelling force than mere instinct has rarely been addressed. Perhaps we humans would be uncomfortable with the repercussions of this news. It would mean that instead of being trapped in preordained behavior patterns, animals are able to choose actions that absolutely override their own species' predilections. In other words, an individual animal can act with free will—and not only free will, but can choose to be kinder and more loving than nature's boundaries normally permit.

Animals can choose to be kinder than even nature intended.

> *"Sentimentalist" is the abuse with which people counter the accusation that they are cruel, thereby implying that to be sentimental is worse than to be cruel, which it isn't.*
>
> —*Brigid Brophy*

William Loyd witnessed a remarkable interspecies rescue in the backyard. A blackbird was harassing a baby rabbit, pecking his head. The rabbit was too young and small to escape. To William's amazement, a pair of robins flew to the rescue, chasing the blackbird away.

These birds went out of their way to protect a being who is obviously not even remotely of their own species, battling an aggressor who is much closer to their species. I often read scholarly arguments that animals are not really being kind when they come to the aid of each other—they are merely attempting to protect their own or similar genetic material. When, for example, a chimpanzee helps another chimpanzee, this is construed as an instinct to ensure that beings with similar DNA remain alive and produce offspring.

This argument has always seemed to me an overwrought attempt to avoid acknowledging the role of compassion in the animal kingdom. Apparently, the thought that kindness exists everywhere makes certain scientists extremely uncomfortable.

William Loyd's story particularly delights me because the robin's genetic material is much closer to that of a blackbird than that of a rabbit. We are left to conclude that the robins were helpful purely because they felt inspired to be kind to the baby rabbit; they clearly saw that it was the right thing to do.

A TINY SPARROW'S HEART

A sparrow asked me for some crumbs
One summer day

I fed her with my trembling hand
She asked me to look away
I looked away and saw my glance
Too fierce for her
So harsh from struggling to live
That gentle creature
Would sit upon my hand I know
If she was sure
I would not crush her with my love
If my love could be that pure.
Then I could kiss a hummingbird
Or stretch its wing
I could watch the fairies dance
On sunbeams in a ring.
If I could be a sparrow's friend
It would be a start
I gave her crumbs, she gave me love
From a tiny sparrow's heart.

—*Jeffrey Armstrong*

Mongooses are small animals, easy prey to many larger carnivores. A mongoose instinctually gets as high up as it can on one of the mounds in its natural habitat. Height and the ability to see danger often means life or death to a mongoose and his tribe. So when a clan of mongooses living in an artificial compound with only boxes to climb chose to sleep on the floor to give comfort to another mongoose who had become too ill to climb, they were disobeying their strongest survival instincts. This is not survival of the fittest. This is compassion for the least fit and, indeed, compassion strong enough to put one's personal safety at risk.

In her book *Mongoose Watch,* Ann Rasa describes how a wild mongoose named Tatu became too wounded to walk. In consideration, the other mongooses would change the direction of their foraging and slow their search for food so that Tatu could keep up and obtain bits of food that they dropped. Tatu was even permitted to eat the same prey another mongoose had caught, which is strictly taboo in mongoose society. When Tatu could no longer groom herself, other members of her clan did it for her.

Mongooses are not like monkeys; they do not simply enjoy grooming each other, and they never groom adults. So the other mongooses were not grooming Tatu to achieve personal satisfaction but out of purely altruistic consideration. As Tatu approached death, the whole clan gave up hunting for food for a few days, instead choosing hunger and staying within the confines of their camp to be as close to her as possible while she died.

"No animal," Rasa said, "takes care of its sick, with one notable exception—humankind—and, as I was able to see with my own eyes, the Dwarf mongoose!"

Little things that run, and quail,
And die, in silence and despair!
Little things, that fight, and fail,
And fall, on sea, and earth, and air!
All trapped and frightened little things,
The mouse, the coney, hear our prayer!
As we forgive those done to us,
The lamb, the linnet, and the hare
Forgive us all our trespasses,
Little creatures, everywhere!

—James Stephens

A little female chimpanzee who lived in a cage at a primate center became mortally ill. She spent most of her time lying on the floor, listless and pathetic, according to the researchers.

The behavior of the other chimpanzees in her compound changed drastically when they were near her. Although usually rambunctious in their play, the other monkeys would scrupulously avoid disturbing the helpless little invalid. They would grab and frolic with each other but were careful to avoid jostling her. Occasionally one would go over to her and touch her tenderly.

They were so committed to keeping her environment calm that if the play of the others became too rough for a chimp he had only to go and sit by her for respite. The others would then leave him alone for fear of disturbing her.

Perhaps as we study more closely and come to revere the animal kingdom, we will see that many animals do care for each other out of genuine concern and that humanity is not a glorious exception to the rule of general selfishness. Whenever we reach out to help another, we are simply acting as do our fellow creatures.

Men will be just to men when they are kind to animals.
—Henry Bergh

Scientists studying vervet behavior were surprised to learn that the tiny monkeys have different meanings for their calls. Chattering signals "snake," so upon hearing another vervet chatter, the group will stand as tall as possible and scan the grass for snakes. A bark means a leopard has been sighted, which sends all the little monkeys running toward the trees. A particular chirp means an eagle has been sighted. Even more amazingly, younger vervets do not know the meanings of the sounds and must be taught. Behaviorists concluded that the vervet language is learned, not instinctive.

Everyone knows that porpoises and dolphins possess a sophisticated sonar-based language and are extremely intelligent. Just how intelligent, we might well have underestimated.

A few years ago, several newly captured porpoises were put into a holding tank at Miami's Seaquarium. Adjacent to this tank was another one in which trained porpoises were kept. The porpoises in the two tanks called to each other back and forth throughout the night.

The next morning, the surprised trainer discovered that the new porpoises already knew how to do the tricks, presumably having been told what to do by the experienced porpoises.

It has been said that the common rat will prove or disprove humanity's own goodness because the rat tests our own compassion. It is easy to love the attractive animals—the cute, fluffy fellows, or those we have seen from childhood in Disney cartoons. But it is the less pretty—those who feel pain and suffer and cannot cry out to us in ways we can understand—that truly test the breadth of our compassion. To these unpopular creatures we must be kind, not out of self-indulgence or the entertainment value of personified traits, but simply because these creatures, too, can suffer, and we need to prevent that suffering.

Being kind to a rat saved the life of a nineteenth-century clergyman. A rat took up residence in his home. The clergyman, noting that it seemed friendly and almost tame, decided not to interfere with its existence.

One night the man was awakened when the rat bit him on the cheek. He looked around and realized his house was on fire. Narrowly escaping, he got out just in time to see his entire home become engulfed in flames and burn to the ground.

Sailors say rats are the first to abandon ship when they sense a serious problem. But this rat returned the clergyman's favor by first taking time from his own flight to save his friend's life.

A West Virginia coal miner noticed a rat watching him carefully as he dug for coal deep underground. It must get terribly lonely down in the mines, because this miner thought the rat seemed friendly and took such a liking to him that he started sharing his lunch everyday with his friend.

Coal mining is arduous, dangerous work and friends must look out for each other. Whenever the miner was about to detonate the explosives that would enable him to dig deeper, he would gently chase the rat out of harm's way.

One day, while the miner worked a vein in an underground room, the rat seemed extremely nervous and ran about in a frenzied manner. The miner decided to follow his rodent friend to see what was wrong. As soon as the two had emerged from the underground room, the ceiling collapsed, precisely where the miner had been standing. He would have been killed had not the little rat run back and forth, begging him to follow.

Sometimes to look more closely is to love:

> *Lightless, unholy, eldritch thing,*
> *Whose murky and erratic wing*
> *Swoops so sickeningly, and whose*
> *Aspect to the female Muse*
> *Is a demon's, made of stuff*
> *Like tattered, sooty, waterproof,*
> *Looking dirty, clammy, cold*
> *Wicked, poisonous, and old:*
> *I have maligned thee! for the Cat*
> *Lately caught a little bat,*
> *Seized it softly, bore it in.*
> *On the carpet dark as sin*
> *In the lamplight, painfully*
> *It leaped about and could not fly.*

Even fear must yield to love,
And Pity makes the depth to move.
Though sick with horror, I must stoop,
Grasp it gently, take it up,
And carry it, and place it where
It could resume the twilight air.
Strange revelation! Warm as milk,
Clean as a flower, smooth as silk!
Oh what a piteous face appears,
What great fine thin translucent ears!
What chestnut down and crapey wings,
Finer than any lady's things
And O a little one that clings!
Warm, clean, and lovely, though not fair,
And burdened with a mother's care:
Go hunt the hurtful fly and bear
My blessing to your kind in air.
—Ruth Pitter

If kindness can be found even in a little rat, might it not be found everywhere? Surely if even rats respond to love, we must live in a world more mystical, more loving, more beautiful, than we have so far conceived. If we think the world a somber place and never take the chance to love the beings in it as completely as we can, we might find our harsh judgment a self-fulfilling prophecy.

> *There may be no greater example of the withholding of compassion to a whole species than our present treatment of the domesticated rat.*
>
> —*Daniel Van Arsdale*

Some people think that compassion in animals could be only an aberration, that it happens so infrequently that this is reason enough to discount its importance. On the contrary, the kindness of animals is so prevalent that it can be counted upon. So reliable is the kindness of one pelican to another, for instance, that it has even been used by humans to their own rather wicked advantage.

Citing the *History of Mexico,* George Romanes tells of men who used the natural compassion of pelicans to get fish for themselves. They would capture, wound, and tie down a pelican, who would then make pitiful noises as he desperately struggled for freedom. Other pelicans, hearing their comrade's cries, could be relied upon to come to help. When they found that they could not free their victimized friend, they would disgorge the fish they had caught in their pouches to feed him. The men would then steal the fish, leaving only a little bit for the captured pelican.

I am the voice of the voiceless;
Through me the dumb shall speak,
Till the deaf world's ear be made to hear
The wrongs of the wordless weak.
And I am my brother's keeper,
And I will fight his fight;
And speak the word for beast and bird
Till the world shall set things right.

—*Ella Wheeler Wilcox*

Because the South American viscachas, a kind of subterranean rodent, often destroy their crops, farmers routinely plug up the underground holes in which they live. But other viscachas will immediately rush out and unplug the holes again to save their trapped comrades.

Man considers a pest any animal that doesn't put money into his pocket.

—Dr. Douglas Latto

Beyond Violence

Standing before you as the advocate of the lower races, I declare what I believe cannot be gainsaid...that just so soon and so far as we pour into all our schools the songs, the poems, and literature of mercy toward these lower creatures, just so soon and so far shall we reach the roots not only of cruelty, but of crime.

—George T. Angell

In coming to grips with the epidemic of violence in American society, educators and social workers are rediscovering the tremendous healing potential of animals.

In response to rampant violence on our streets and in our schools, our society is waking up to the importance of teaching the way of kindness to our children. But developing this necessary compassion is a spiritual art. Counselors have found that because domestic animals are sweet and nonthreatening, children often find it easier to empathize with them than with other children. Advertisers learned long ago that they can sell anything to children by using animals. But animals are not simply fascinating images—they are living spiritual beings whose purity and directness communicate to children outside the complex webs of adult compromise and cynicism.

Liz Helms of Ahead with Horses works with juvenile delinquents. She is deeply impressed by the healing that occurs when young offenders take care of horses. They bear a great deal of responsibility, she said. If, however, one of the animals makes a mistake, or goes lame, or cannot function one day, no one stops loving it: no one wants to lock it away or punish it. Such forgiveness is often something new for the delinquents who, prior to this, had never received a second chance from others or, most of all, extended one even to themselves.

Given the apocalyptic wars and environmental destruction resulting from the philosophy that earth and animals are merely things to be owned and used as man sees fit, the world might now be ready to recognize the sanctity of the spirit in animals and throughout Nature.

Around the world, humanity is awakening to our need to reharmonize with nature, to find a way to cooperate with the life-system of the planet, rather than trying to bully it into submission. In the face of environmental and social collapse, all our ideas about society, technology, and lifestyle are necessarily changing. In the past decade, we have witnessed a worldwide spiritual quickening that has altered our attitudes to animals.

This quickening of global conscience brings us to a renewed appreciation for the value of aboriginal wisdom. The native peoples of North America honored the animal kingdom from the heart, realizing that our brotherhood with animals nourishes our human relationships.

"The old Lakota was wise," wrote Chief Luther Standing Bear. "He knew that Man's heart away from nature becomes hard; he knew that lack of respect for growing, living things soon led to lack of respect for humans too."

Our relationship to animals is perhaps the most direct, meaningful way in which we can encounter the Other in Nature. Animals relate to us with an immediacy, a presence that conveys equality, respect, and goodwill. As we recognize the soul in animals, and accord them the full respect they deserve, we will be empowered to make the greater changes that can save our planet.

> *Show me the enforced laws of a State for the Prevention of Cruelty to Animals and I in turn will give you a correct estimate of the refinement, enlightenment, integrity, and equity of that Commonwealth's people.... The lack of humane education is the principle cause of crime.*
>
> *—Hon. L. T. Danshiell*

The violence that plagues our society has long been linked with incidence of cruelty to animals. In fact, childhood exposure to animal abuse is viewed as an indicator of domestic violence, child abuse, assault, rape, and murder.

"Animal advocates have known for years that sometimes abuse of animals can be a warning sign," says Martha Armstrong of the Humane Society of the United States. The Humane Society is developing a national standard for reporting violence toward animals. They maintain a database to assess the pervasiveness of this problem and develop intervention programs.

Political leaders are just beginning to see the connection between kindness to animals and healing the epidemic of violence in our schools. "Just as soon and so far as we put into all our schools more humane education," said former Governor of Texas Miriam Ferguson, echoing George T. Angell, "and foster the spirit of justice and kindness towards the lower creatures—just so soon and so far shall we reach the roots, not only of cruelty but of crime."

> *If you have men who will exclude any of God's creatures from the shelter of compassion and pity, you will have men who will deal likewise with their fellow men.*
>
> —*St. Francis of Assisi*

If one allows the possibility that there can be animal angels, then one must allow that there is a spirit of goodness that runs through all life. If kindness can be found in lions, rats, monkeys, pigs, ants, birds, earthworms, and South American rodents, then surely it is everywhere.

To me, this goodness is the only real antidote we have for violence. We will not achieve social harmony by locking up our brothers and sisters in cages, whether we do this by enclosing animals in zoos or enclosing our own kind in prisons. Society is more likely to improve if we expand and nurture with love the good or, at least, the natural character of every precious being in God's creation. And that includes our own.

Recently a young boy named Luke Woodham opened fire at his school, killing several schoolmates and a teacher. He had murdered his dog shortly before. A couple of months later an even younger child shot and killed several of his classmates. I checked the papers to see if my suspicions were correct. Indeed, he had gone hunting for the first time the week before he killed at school, and he had killed an animal.

Benjamin Franklin observed: "We must all hang together or surely we will all hang separately." This is true of life on earth now. We either begin to respect and love and nurture all life, or we fail. It's that simple. We either teach our children love and concern and respect for other beings, or we don't. We either save the animals, the environment, and the planet, or we fail.

This takes commitment. But we do have a secret weapon on our side. Love and kindness are woven into the fabric of the beingness of all beings, our own as well. And nowhere is this more plain to see than in the natural love that children have for animals.

Kids love animals and they need to know that kindness is a natural part of existence. Showing children that animals are intrinsically capable of kindness enables teachers to teach kindness without appearing preachy. Such lessons also demonstrate the continuity of goodness and compassion throughout all life.

Students in one class wrote essays about their animals (mostly dogs and cats), and I want to share some of the things the children wrote:

> He can't stop loving people so much. He wants to sit by people.
>
> If we didn't have animals on earth we would be lonely.

We must protect the animals. When I grow up I am going to protect all the animals.

Sometimes I feel good inside thinking about animals.

Some people don't like animals. I love every animal in the world.

I have a turtle. Her name is Selena. She is like my child.

When I grow up I am going to visit the President and tell him to stop the hunters from killing birds.

April [her bird] keeps me company by singing when I do my homework.

I like the birds because they make the day happy.

Whoever has a pet should be thankful and proud to have a pet. Treat it like a real person only it looks different.

Even if animals bite, they are still my friends because they don't know any better.

I love animals more than anything else in the world.

This reminds me that the second part of that quotation in the Bible about the lion lying down with the lamb is: "And a little child shall lead them."

> *We have become nuclear giants but ethical infants; we know more about war than we know about peace, more about killing than we know about living.*
>
> *—Gen. Omar Bradley*

Canned hunts are set-up hunting expeditions for wealthy tourists who pay large amounts of money for the thrill of killing a ferocious animal and taking their trophy home with them. However, most of these tourists are amateurs who do not want to have to confront the animal on fair terms (not that using a gun against any animal is ever fair).

To ensure the tourists' hunting success, the hunt organizers first drug the animals so that they never have a chance. The dazed animal just sits placidly while the tourist creeps up and fires at close range without risk. According to a news story from the United Kingdom, one canned hunt took a hilarious twist.

A wealthy American hunter was willing to pay well for a wild bear hunt in Russia. But wild bears are difficult to come by near Moscow, so the organizers purchased an old bear from a Russian circus and set it loose in the Perdelkino Forest outside Moscow.

The hunter was similarly set loose with some encouragement as to which direction to go and soon found his bear. But before he could shoot, a bicyclist came riding along the path through the forest. The bicyclist was so startled to see a bear he crashed his bike and ran off into the woods on foot. At this point, the bear, having been trained with bikes in the circus, promptly boarded the bicycle and began to ride it around.

The hunter, now furious that he had been deceived with a trained bear, took off without firing a shot.

> *You ask people why they have a deer's head on the wall. They always say, because it's such a beautiful animal. I think my mother's attractive, but I have photographs of her.*
>
> —*Ellen DeGeneres*

A hunter traveling through Africa decided to shoot a gorilla. His guide enabled the hunter to get very close to where the gorillas were resting in a tree. The hunter took aim and shot a young animal, who tumbled to the ground.

The young gorilla was dead, but before anyone could react, an old gorilla rushed to the scene, grabbed the body of the young gorilla, and ran with it into the jungle, thus thwarting the hunter from claiming his quarry. The hunter was astonished by this scene, so like when a bereaved human recovers the body of a slain loved one.

Money, unlike animals, can talk.
—Anonymous

Angels in Hollywood

Many Hollywood trainers, especially in the old days, used cruelty and coercion to control large animals like lions and tigers. One of the first people to give up this practice and raise even big cats and other large animals with kindness is a man named Ralph Helfer, who called his program "affection training."

Raising animals on a strict diet of love, Ralph produced animal actors who genuinely cared about him and were part of his family. At first, critics scoffed that this was impossible to do, but Ralph eventually developed a reputation as a trainer whose animals could work even with small children. Ralph believed that love and kindness could be used to train even the most ferocious animals and in the most desperate circumstances. His theory was put to the ultimate test one day on the Hollywood Freeway.

A beautiful lioness named Tammy was traveling in a trailer especially designed for her to a studio in which she was scheduled to begin filming. But the trailer-hitch on the car broke at the worst possible time—two freeways had just merged and now the car and the broken-hitched trailer were forced into the fast lane as the roads combined. The metal snapped and Tammy's trailer turned over on its side, shattering its walls.

Tammy climbed out of the wreckage. A fully grown African lion was loose on the Hollywood Freeway. She was wounded, and she had blood in her eyes so that she had trouble seeing. Tammy wandered in a daze as cars sped by. Ralph got out of the car about fifty yards from the injured and confused lion, took a deep breath, and went into action.

If ever affection training was to be proven, it was now. Ralph sweetly called to Tammy and watched anxiously to see what she would do. Would the lion revert to fear in this strange and terrifying situation, or would she follow the voice she had come to trust through the noisy, chaotic traffic back to her friend?

Ralph called again to her and she wobbled forward uncertainly. Cars stopped, as motorists watched the drama in disbelief. If Tammy didn't respond, she would surely be killed, or even worse, she might kill others if she panicked and ran across the freeway to the residential area beyond.

Tammy slowly staggered toward Ralph. Time seemed to stand still as she made her way back to her friend. With only her trust to guide her, she put aside all her instincts to bolt and walked steadily toward the voice she knew and loved.

When she got close to him she began making the throaty chuffing sound lions use for friends. In *The Beauty of the Beasts*, Ralph describes the scene this way: "A lump was forming in my throat. Here was this blind lion on the freeway, battered and nearly unconscious, limping toward the sound of my voice. To this day, I have never been more proud of our affection training."

Tammy was taken in the car to the veterinary hospital where she fully recovered.

Ralph Helfer's dramatic success with affection training made possible groundbreaking films like *The Lion,* in which actors could safely work with Zamba, a real lion who had been affection-trained. The importance of the shift to affection training in the motion picture industry was underlined at the 70th Academy Awards. Bart, the huge grizzly bear who had starred in the Academy Award-winning film *The Bear,* made an appearance with his trainer, a man who bets his life everyday on the truth and beauty of affection-based training.

A friend of mine, whose last name, ironically, is Lyons, lived in Africa and raised big cats by using affection training. He told me of a time when he was wrestling with his lion buddy and the lion accidentally knocked him out. He awakened to the abashed lion trying to revive him by licking his face. However, because lion tongues are used for ripping raw meat off bones, it felt like a cheese grater on his face—almost unbearable. He made certain they never played that roughly again. Once, when he had just bought his new bride from America, his lion was so glad to see him that he greeted them in the morning by jumping on their bed with such enthusiasm that the bed broke.

Desmond Morris once met a group of chimpanzees who acted in films. They liked watching TV and would change channels frequently to find the noisiest Westerns.

Today, major actors and directors simply will not tolerate any mistreatment of animals for the sake of a shot. In 1980, James Mason refused to play opposite Sophia Loren in a film that contained a real cock-fighting sequence. "I don't think you should hurt or kill animals just to entertain an audience," he said. "Animals should have some rights. But there are a lot of directors, including Ingmar Bergman, who will injure animals to further a plot. I will have none of it."

Margaux Hemingway was making a documentary about her famous grandfather. The culmination was to be a bullfight. Margaux had never seen a bullfight before and had been told it would be a grand display of man versus nature. However, the carnage did not live up to that description; she saw only a poor animal die an awful and bloody death. Her natural goodness caused her to lose faith in the project.

God gave his creatures light and air
And water, open to the skies;
Man locks him in a stifling lair
And wonders why his brother dies.

—Oliver Wendell Holmes

In *The Lady and Her Tiger,* Pat Derby, a Hollywood animal trainer, describes how she used to care for wolf cubs. She discovered that happy wolf cubs actually sing!

To feed her rambunctious, hungry charges, Pat kept track of them by putting them all in a box labeled "not fed" and then transferring them one by one after their dinner to a box marked "fed." As each fat little wolfling reached the "fed" box, he would start to make a blissful little humming sound until the whole box was full of happy, humming little wolf pups.

Gordon Haber, an expert on wolf behavior, says: "To harvest one or a few wolves from a pack can have a drastic effect on the entire pack. That wolf fur in one's parka or the wolf hide hanging on one's wall will have brought about every bit as much disruption and grief in some wolf pack as would the death of a loved one in a human family."

While I was answering questions on a radio show, a caller related this story of an encounter with wolves that upsets all our expectations about their behavior.

The man had been camping in Alaska and had to return to town for supplies. He left his wife alone at camp.

When he returned, he was shocked to find wolves circling the campsite, snarling at him when he tried to get closer. He called out his wife's name, dreading to discover what had become of her. To his relief, she suddenly appeared—safe and well.

But the wolves remained between them and would not let him approach. Then he realized that, rather than threatening her, the wolves had been *guarding* her. Only when she reassured the wolves that he was a friend did they allow him to approach and rejoin his wife.

> *The man who is described as behaving like a beast would often in his behavior be a disgrace to any known animal.*
>
> *—Ernst Bell*

Wondrous Intuition

Animals can sometimes sense danger where we cannot. When I met with scientist and author Patrick Flanagan, he confirmed this remarkable story of how his dog's clairvoyance saved his life.

Patrick and his wife Gael were driving along a narrow mountain road cut into a very steep cliff. Suddenly, their little white dog, Pleiades, started barking and causing a terrible commotion. Patrick and Gael instantly took heed and slowed their car to see what was the matter. As they crept around a blind curve, they saw that an oncoming car had completely blocked the road. The driver had unwisely picked that spot to do a U-turn, blocking both lanes.

Because he had slowed in response to Pleiades' warning, Patrick was able to stop his car before it slammed into the car blocking the road. Without Pleiades' marvelous knowledge that danger lay beyond the turn in the road, Patrick and Gael would have careened over the 100-foot drop of the cliff and been killed.

We are challenged to explain how animals can comprehend the nature of a remote crisis without attributing to them a kind of extrasensory perception.

Old Smoke, a normally docile and well-behaved horse at the Guadalupe Ranch in Arizona, suddenly began acting very strangely. He reared his head and galloped about erratically. The ranch hands watched in amazement as Old Smoke bucked so ferociously that he finally broke down a section of the corral and escaped.

The men jumped on their own horses and chased him, but Old Smoke was too fast for them and seemed to know just where he was going. Following the crazed horse as fast as they could, the ranch hands came to an isolated spot in the desert seven miles away.

There they found Kelvin Jones, a guest at their ranch, lying on the ground. The horse Kelvin had been riding had bolted and thrown him. Kelvin had badly injured his ankle and without sufficient food or water would have been in serious trouble if Old Smoke had not come after him.

Kelvin was taken back to town for medical treatment. Old Smoke went from being thought crazy to being praised as a hero. That evening, the heroic horse was treated to a big bag of apples. Old Smoke had been the first horse Kelvin had ridden when he came to the ranch; evidently a bond of friendship had grown between the two, a bond that had been more than physical. When Kelvin needed him, Old Smoke broke all the rules to be with him.

Animals don't feel as we do, we are often told, without ever considering how we can tell even what another *person* is feeling. Science attempts to mechanize and measure everything, yet it has never been able to directly measure the power of love. Instead, some scientists have used their talents to measure things that should, perhaps, be best left unmeasured. But our thoughts do affect each other, even across distances.

In one horrific experiment, Soviet scientists took the babies away from a mother rabbit and put them on a submarine which then submerged. The mother rabbit was connected to an electronic device, not unlike a lie detector, that measured her impulses. One by one, the young rabbits were killed in the submarine. Each time one of her babies were killed, the mother reacted instantly. A mother's heart can sense her children even across great distances.

> *It is an established fact that the training of the intellect alone is not sufficient. Nothing in this world can be truer than that the training of the head, without the training of the heart, simply increases one's powers for evil.*
>
> —*Ralph Waldo Trine*

In a fascinating experiment, human beings tried to inhibit the growth of the fungus *Rhizoctonia solani* from a distance of one-and-a-half yards by using only their thoughts. After incubating the fungus again, its growth was shown to be significantly hindered. In fact, the growth was measurably impaired in 151 out of 195 batches.

What is interesting about this experiment is that the results are clearly readable. Fungi either grow or don't, and the rate of growth can be precisely measured. But what if the human beings were placed one mile, two miles, or three miles away from the fungus cultures? Could they still project their discouraging thoughts successfully?

Researchers placed people at distances of one to fifteen miles away from the different cultures. In all cases they were able to retard growth of the cultures. Bacteria responded in the same way.

If just thinking hindering thoughts about fungus and bacteria is able to stop their growth, what does that teach us? Spiritual masters through the ages have warned us against passing judgment on each other. Perhaps these experiments show that we could be retarding the growth of those we know (and even love) when we think limiting thoughts about them.

Apparently, all life forms are connected so profoundly that distance does not matter. In view of the fact that mere limiting thoughts were able to stunt the fungus growth as completely at a distance of fifteen miles as at one mile, we would not be misguided in sending healing and uplifting thoughts to all life forms. Including ourselves.

We are all connected, whether it be in a circle of love or a circle of hate or indifference. But connected we are—even down to the tiniest and seemingly most insignificant fungus and bit of bacteria.

Lady Dowding foiled an invasion of silverfish, a pesky insect, by simply asking them very nicely if they would kindly depart. They did.

> *Failure to recognize our responsibilities to the animal kingdom is the cause of many of the calamities which now beset the nations of the world... nearly all of us have a deep-rooted wish for peace—peace on earth; but we shall never attain the true peace—the peace of love, and not the uneasy equilibrium of fear—until we recognize the place of animals in the scheme of things and treat them accordingly.*
>
> *—Air Chief Marshal Lord Dowding*

The murderer stole into the room. There he spotted his helpless victim. Savagely he tore her apart, ripping and shredding, destroying and mutilating, until there was no life left. But there had been a witness.

This was the beginning of a most unusual experiment on the consciousness that pervades all life. The victim was a plant. The witness was a plant. The murderer was a human being.

Cleve Backster's life had been changed when he discovered accidentally that a plant rigged up to a lie detector-like device will respond to even the *thought* of abuse to it.

For this experiment, Cleve had six men walk back into the scene of the murder, one at a time. When the man who had played the role of murderer walked into the room, the needle connected to the witness plant jumped. It had an emotional reaction to the man who had murdered its companion.

What does this mean in the greater scheme of things? Certainly, it is another sign that all life is connected, that all of life is conscious to some degree. The idea that we are all a part of a great family does not sound quite so far-fetched as it once might have.

Not content with the awe-inspiring implications of this experiment, Cleve Backster went further. He threw brine shrimp into boiling water; the plants in the next room reacted. He cracked the shells of fertilized chicken eggs and the plants reacted again. In view of this data we might well ask: What happens to the plants on a battlefield when blood is shed?

The Gaia theory views Planet Earth as one giant conscious being who can react to the things that happen to her, to the suffering her children feel, and to the love they create. If it can be shown that plants react sympathetically when brine shrimp are boiled alive, then might we extrapolate that Mother Earth herself does indeed weep when her children destroy her body and cause one another pain?

And how long shall we continue to go on pretending that this is unimportant?

LIFE

For God's sake, kill not: Spirit that is breath
With Life the earth's gray dust irradiates;
That which has neither part nor lot with death
Deep in the smallest rabbit's heart vibrates.
Of God we know naught, save three acts of will:
Love that vibrates in every breathing form,
Truth that looks out over the windowsill
And Love that is calling us home out of the storm.

—Eva Gore-Booth

"Surely you will be getting rid of your cat now that you're having a baby" is something that pregnant women often hear. But Kandy Phillips is glad she never listened to such advice.

Kandy was doing her dishes when her cat Oscar began meowing loudly at her. When she ignored him, Oscar jumped up on the kitchen counter, something he had been taught never to do. Kandy brushed him off. Like a bad kitty, he jumped right back up again, and when she pushed him off a second time he bit her firmly on the leg. Then he started running about and meowing.

Following her disobedient and strangely behaving cat, Kandy walked into the bedroom where her baby, Anthony, was sleeping in his crib. The cat jumped onto the crib.

To her horror, Kandy saw that little Anthony had turned purple and was not breathing. A thin stream of vomit by the side of his mouth showed that he had coughed up in his sleep and choked. Kandy rapped him on his back and attempted resuscitation. It didn't work. Finally, in desperation, she hit him soundly on the back, whereupon he cried out and coughed up more vomit. He made a gasping sound as his lungs filled back up with air.

Oscar had brought Kandy to the crib just in time.

The animals of the planet are in desperate peril. Without free animal life I believe we will lose the spiritual equivalent of oxygen.
—Alice Walker

One cannot research stories of animal angels without being struck by the many accounts of animals that have come back from beyond the veil to visit a loved one. Normally I do not include such stories, but this one was told to me by a very sincere person while I was on my book tour and I found it charming.

Olivia had white fur, blue eyes, and a little pink collar, and she was considered by those who knew her to be the world's best cat. People even asked about her on the answering machine—that's how well loved she was by everyone who met her. Part of Olivia's charm was the way she would run up to anyone who came in, purring and issuing little yelps of gladness.

One day, however, she greeted a car with a little too much eagerness and her life was over. It was weeks before the depression began to lift from the neighborhood. Even the other two cats in the household seemed lost—they hated each other, but loved Olivia to the point where they'd actually step back from the food bowl and let her eat first. Sadly, they took to hanging out in her old favorite spots, interminably searching as though they might find her somewhere.

One night after Olivia had been dead for some time, Olivia's person looked up from her reading to see Nell, her other cat, standing outside the window. Nell didn't seem to be trying to attract her attention, so she continued to read. Suddenly she heard this great *woompf*, as though the window was going to cave in. She got up and went to the window, hoping by her stern expression to convince Nell to be a bit more patient and not throw herself against the glass.

To her shock, Nell was no longer there. Instead, there sat a little white cat with blue eyes and a bright pink collar. No, it can't be, she said to herself over and over, but she kept looking, and it was, without a doubt, Olivia, looking at her sternly. She felt Olivia was admonishing her for not treating her other cat, Olivia's sister, with the same lovingness that came naturally with Olivia. And she felt thrilled, hoping against hope that the little white cat buried in the garden was somehow not Olivia. But when she ran to the back door to let her in, there was nothing there.

Her husband said later that the loud noise was the sound of a cat so spoiled that they threw her out of heaven, landing on the patio steps. "But," she writes, "I think Olivia wanted to give me one more chance to remember her as she was, instead of as I saw her when I buried her under her favorite dwarf maple."

> *That the Bible gives any ground for the general fancy that at death an animal ceases to exist, is but the merest assumption. Neither is there a single scientific argument, so far as I know, against the continued existence of animals, which would not tell equally against human immortality. While I believe for myself, I must hope for them.*
>
> —*Rt. Rev. George MacDonald*

Religious people believe that throughout history angels have been a solace to humankind. They have provided inspiration, a balm to the soul, helped us when times were most desperate, and cheered us on when we became despondent. These are the very ways in which animals help us every day.

It is not my task to dispute or purport visitations from heavenly beings. I merely note that a very large number of beings act in ways we might consider unusually loving. Perhaps, as well as looking toward higher beings for our inspiration, we might do well to look within for the best parts of each other. The angels might be closer to us than we think.

Trial by Fire

Kathie Vaughan was driving the used truck she had purchased that morning when it suddenly began to fishtail. She finally managed to bring the vehicle to a stop with a loud screech.

But her troubles had just begun. The cabin interior was filling fast with noxious fumes and black smoke. Most people could have simply jumped out of the vehicle, but Kathie is a paraplegic—paralyzed with multiple sclerosis from the waist down. Her truck was on fire and she knew she could be blown up at any moment.

Kathie shoved her Rottweiler Eve out of the door, along with her wheelchair. But due to the thick black smoke, she could not find the wheels to the chair. A throbbing panic overwhelmed her. She had to get out immediately, before the truck exploded.

In danger of blacking out, Kathie suddenly felt Eve, refusing to desert her, grab her leg with her jaws. Eve firmly grasped Kathie by the ankle and dragged her ten feet to relative safety. Then the truck burst into flames. Ignoring the terrifying fire, Eve dragged Kathie to a nearby ditch.

A police car arrived on the scene. "You've got to get further away!" the policeman shouted to Kathie. The truck was on fire, the flames were approaching the gas tank; there was danger of a terrific explosion.

Kathie struggled to pull herself away from the truck and towards the police car. Her head and body ached with pain and she found she could hardly move. Eve bent close to her human friend, offering Kathy her collar. Then Kathie held tight as the determined dog dragged her forty feet to safety.

The firemen eventually extinguished the blaze. Eve was awarded the prestigious Stillman Award for bravery.

The car careened out of control. Sixty-year-old Jesus Martinez was having a heart attack at the wheel of his car and could no longer control the vehicle. The other drivers on the Houston, Texas, road had no way of knowing that their lives were suddenly in danger, too: Jesus was too far gone to stop the car.

Also in the car, Jesus' Schnauzer Bitsy realized that something was terribly wrong. Whether it was intelligence or instinct or just luck no one will ever know, but Bitsy threw himself against the steering wheel in such a way that the car turned onto the shoulder, out of the way of the other cars.

But the car was still going too fast. So Bitsy bit Martinez, causing him to take his foot off the accelerator.

Martinez was taken to the hospital and, thanks to his dog's actions, later recovered.

> *Anyone who has accustomed himself to regard the life of any living creature as worthless, is in danger of arriving also at the idea of worthless human lives.*
>
> *—Albert Schweitzer*

Nothing seems quite so human as talking to yourself. That is why it is both delightful and unsettling to learn that primates who have been taught sign language also talk to themselves.

Washoe was one of the first chimpanzees to learn American Sign Language. Researchers noticed that when she was signing to herself, she would pull away as if she did not wish to be observed. Sometimes she would take a magazine and climb up a tree so she could sign to herself and be completely alone. In one study of 5,200 chimpanzee conversations, 119 of the signed conversations were private.

> *Besides love and sympathy, animals exhibit other qualities connected with the social instincts, which in us would be called moral; and I agree—that dogs possess something very like a conscience.*
>
> *—Charles Darwin*

In *Walking with the Great Apes*, Sy Montgomery states that some of the material gathered in observing the Gombe chimps has never been published, simply because the findings seemed too fantastic to be believable.

Two naturalists were talking to each other when one of them mentioned he wanted to show the other something incredible.

Reminding his friend to keep quiet, the first man walked over to a chimp named Fifi, who had injured her foot. Talking in a monotone voice he said, "Blah, blah, blah, blah, and Fifi, show me your foot." At once, Fifi held her foot out for him to see. He repeated this exercise, speaking nonsense and then suddenly asking Fifi to show him her foot; Fifi again responded to his words at the appropriate moment. This experiment was witnessed by five other people.

Many people have claimed for years that their animals understand English. While such accounts do not constitute incontrovertible proof, it seems to me that some of the most interesting data may be that which goes unreported because of the prejudices of the human race.

Very well documented, however, are the studies of Koko the Gorilla, who can currently demonstrate her understanding of about 2,000 English words in American Sign Language.

When Koko was asked how she spends her birthdays, she replied in sign language: Eat, drink, [get] old.

Alex, the famous African gray parrot, also seems to *understand* as well as talk. Professor Irene Pepperberg asked Alex how a green bottle and a green hat were similar.

Alex replied, color.

When asked how the objects differed, Alex responded, shape.

In fact, Alex can identify five shapes, seven colors, and fifty objects. He was looking at himself in the mirror one day and asked a nearby person, Color? Now, when asked what color he is, he replies that he is gray.

> *If there is anything in Carl Gustav Jung's concept of a universal consciousness, the combined outrage of the millions of creatures which have suffered at the hands of man may well combine to haunt us. We are all of the same family, though destiny has assigned us to different roles: in our relationship with animals, we should regard them as different, not inferior.*
>
> —*Dennis Bardens*

Alfredo felt himself begin to lose his balance. The eleven-year-old boy had been playing on the roof of a new house that was under construction near his home in Salerno, Italy when he lost his footing. Down he tumbled, thirty-five feet, knowing that only hard concrete would break his fall. Screaming, he wondered if he would be paralyzed or even killed.

His dog, Stella, a big German Shepherd mix, had been watching Alfredo anxiously. When the boy started to fall, Stella shot into action. Running as fast as she could, she placed her body where Alfredo would land.

He later said that when he hit it was like falling onto a mattress. Both Alfredo and Stella were fine.

Kyle Leibach was trying to get a good night's sleep but his cat Delores just wouldn't let him. Again and again she jumped on the man's face and scratched him. Kyle kept thinking this was unusual, as Delores was normally a very quiet and shy cat. Again he pushed her away.

Then Kyle noticed something was very different. He usually left all the lights on in his apartment because it seemed to help Delores to feel more secure, but now the apartment was black—black with smoke.

He immediately began to panic; he could hardly breathe. The windows all had bars over them. The only way out was by the back door and its doorknob had recently broken. Kyle turned it and it fell off and landed at his feet.

With his only way out blocked and panic overtaking him, Kyle lost consciousness, due to lack of oxygen. But Delores would not let him die. She scratched his face again until he regained consciousness. Kyle stood up, threw his whole weight against the door, and broke it down.

Delores was still trapped inside. She had lost consciousness. Kyle was greatly relieved when a fireman found her. For an hour they worked on the brave little cat, administering CPR and oxygen to her. Her eyes had been seared shut from the flames, she couldn't eat and could barely drink water.

But Delores fought hard for her own life as well. On the fourth day she regained her strength and is now recovered.

> *No one really needs a mink coat in this world...except minks.*
>
> —*Glenda Jackson*

We are so used to hearing of noble dogs saving children and entire families, that it may come as a shock that the self-contained cat could also be a hero. Actually, the ordinary housecat often rises to challenges beyond the call of duty. For example, a very brave cat named Lucy put herself between a two-year-old toddler and a rattlesnake that was winding its way down the hallway of a California home.

The child's mother first became aware of the danger when she realized that the hissing sound she heard was not the water sprinkler, as she had first thought, but was coming from inside her house. To her horror, she saw a deadly snake approaching the bedroom just as she heard her child wake up and begin to walk to the hallway. Lucy jumped between the child and the snake and courageously fought it off, saving the life of the child. Lucy herself emerged from the skirmish unhurt.

If you take all your books, lay them under the sun, and let the snow and rain and insects work on them, there will be nothing left. But the Great Spirit has provided you and me with an opportunity for study in nature's university, the forests, the rivers, the mountains, and the animals, which include us.

—*Tatanga Mani*

The sea cow was first discovered by humans in 1741. These incredible animals grew up to thirty feet long and weighed up to four tons. They roamed the oceans in great herds.

Sea cows demonstrated remarkable caring for each other. A naturalist noted that when one sea cow was captured, others would rally around and attempt to save their comrade. The sea cows would begin circling the wounded cow while others attempted to ram the boat containing their friend's tormentors.

One day a female lay dead upon the shore. For two days her mate, despondent and refusing to accept the loss of his dear one, swam closer and closer to shore. He watched and waited and swam and called, but she could never come back to the sea with him again.

Twenty-seven years after man first discovered this gentle animal, the sea cows had been hunted to complete extinction.

> *Let us remember with humility the loneliness of being man in a universe we do not understand and the vulnerability of the human condition. The animals could do very well without us, but we cannot do without them.*
>
> *—Gerald Carson*

Orca babies learn to speak in whale dialects that vary from pod to pod. When two orcas were introduced into the same tank, scientists noticed that they changed the inflections of their speech to learn to talk together.

David Bain of Marine World noted that when a dolphin named Bayou was separated from his friend, an orca named Yaka, Bayou began to call to Yaka using orca-like sounds.

ASCENDING OF THE ANIMALS

The Animals, you say, were sent
For Man's free use and nutriment.
Pray, then, inform me, and be candid,
Why they came aeons before Man did,
To spend long centuries on earth
Awaiting their Devourers' birth?
Those ill-timed chattels sent from Heaven
Were, sure, the maddest gift ere given
Sent for Man's use (can Man believe it?)
When there was no Man to receive it!

—Henry S. Salt

The Souls of Animals

I swear I think now that every living thing without exception has an eternal soul. I swear I think that there is nothing but immortality.
—Walt Whitman

Do animals have souls? Can they experience and share love?

I used to believe in the Darwinian notion that life on earth was all about survival of the fittest and the law of the jungle, because that's what I was taught in school. I now perceive a divine spark in all beings, including those in animal kingdom. The stories in this book show that animals can act with the highest degree of love, valor, compassion, and goodness. These stories are the animals' own communication to the human world, proving that they have souls and must be treated with goodness in return.

It is pointless to dispute overmuch about the term "soul" because no one can absolutely prove that he or she has a soul. We can only infer the presence of soul by the evidence it leaves behind. Scientists prove the existence of certain ephemeral elementary particles by their aftertrails through a bubble chamber. Souls also leave aftertrails through our lives. When we consider a person like Mother Teresa, and contemplate a life spent in utter selfless service, we feel the presence of a powerful soul-force. Just to hear of such a being and to imagine an existence of perfect dedication and selfless love quickens something in ourselves. Like musical instruments whose strings begin to vibrate when a tuning fork plays one of their notes, something in us resonates when we hear of acts of selfless kindness and beauty. This might be the only proof we will know until we leave this earth behind.

If having a soul implies the ability to recognize another sensitive, intelligent being, regardless of its outer form, and to care for and assist another out of selfless compassion, then animals, both tame and wild, reveal their soulful nature to us again and again. The testimony of the animals themselves forces us to acknowledge their soul-nature and hence to reevaluate our entire relationship to the animal kingdom.

If life means anything at all, if we beings are connected in any way spiritually, then somehow whatever we do to the least of us, the weakest and most helpless, we do also to ourselves.

In our treatment of animals, nothing less than the direction of life itself is at stake. If we continue to act as if only our own selves matter, and that the suffering of others matters not, we will continue to create a world of suffering for animals, and we will suffer the psychic energies and imbalance of this decision. We will feel more and more alienated from our own world, not even fully comprehending that it is a world of our own making, instead attempting to find order and peace through drugs or monetary gain. What is precious about life can grow only from honoring *all* life.

Reason often makes mistakes but conscience never does.
—Josh Billings

In *The Souls of Animals,* author Gary Kowalski speculates on the concept of play. He makes the point that we ought to take seriously the fact that play is joyful and fun. "It would be easy to imagine animals training for the demands of survival mechanically and mirthlessly," he states, "without sport or amusement. Yet in fact, anyone who has ever had a puppy or kitten or watched a program about lion cubs, knows animals clearly *love* to play."

Much of our cultural attitude toward animals is derived from the theories of Descartes, who saw the universe as a clock-like mechanism in which only humans possessed souls, and therefore feelings.

Someone should have pointed out to Descartes that animals are not clock-like automatons but feeling beings, and that this proposition is proved by the fact that clocks do not frolic and tease and make merry with each other when left alone. Furthermore, if you leave two clocks alone in a room for awhile, you do not eventually have a lot of little clocks running about.

Dostoyevsky said: "Love animals: God has given them the rudiments of thought and joy untroubled. Do not trouble their joy, do not harass them, do not deprive them of their happiness, do not work against God's intent."

Only because we have been blinded by a rationalism that celebrates the negation of sympathy could we have ever missed the obvious fact that animals are living beings, just as we are, not complex mechanical devices. The difference between a machine and joyous animals frolicking on a hillside—whether they be goats at play or seagulls ecstatically riding updrafts or kittens wrestling and tumbling together or otters sliding down muddy banks again and again out of the sheer joy of being an otter—this difference is so immediately apparent that the idea of animals as automatons would be ludicrous if people hadn't taken Descartes' arguments with the uncritical acceptance of an hypnotic subject.

But this specious argument was taken seriously. According to Gerald Carson in *Men, Beasts, and Gods: A History of Cruelty and Kindness to Animals*, during the time of Descartes it became fashionable to perform experiments on live animals after elegant dinner parties. Puppies were attached to awful mechanical devices and tortured or drowned in the interest of scientific experimentation. The genteel folks performing these experiments told themselves that inasmuch as animals, by definition, could not feel real pain as we do, such studies and experiments were perfectly all right. This soul-blindness continues to wreak agony upon animals today behind the locked doors of modern vivisection laboratories.

Great thinkers have always realized that what we know intuitively may be far more important than what we think we can discern through logic alone. Thoreau saw this clearly when he said, "Our science, so-called, is always more barren and mixed with error than our sympathies." John Vyvyan shared this view when he stated that "Knowledge without pity may well be the greatest danger that besets the world."

When the logicians finally decide whether or not animals have souls, or whether we ourselves have souls, or even figure out what a soul is—we shall nevertheless have to admit the incontrovertible fact that animals, like us, do play, and that the joy of life they obviously feel when they play ought to be reason enough to respect them. Respect for joy itself—the joy that can be found in every living creature—ought to be reason enough to leave them in peace. As Samuel Butler once said, "All animals except man know that the ultimate purpose of life is to enjoy it."

A POEM TO DESCARTES

I think therefore I am, he said
As he cut apart
A puppy's head.
You ought to feel as well, said we.
You've blackened human history.
—Stephanie Laland

Sometimes animals save our lives by simply being there.

When a U.S. naval vessel was torpedoed during World War II, the men rushing to escape in life rafts were careful to grab the ship's cat Maizie and make sure she was safe with them.

The men spent the next forty-five hours drifting and praying they would be picked up by friendly forces. Maizie sat in their laps and ate malted milk tablets with them. She would sit in each man's lap and comfort him, in turn.

After they were rescued one of the sailors said: "If Maizie hadn't been with us we might have gone nuts."

Sue Strong is a disabled woman with a capuchin monkey named Henrietta who has been trained to help Sue with many tasks, such as turning on the lights. Sometimes when Sue is reading, Henrietta will take it upon herself to dim the lights so Sue will have to ask Henrietta to turn them back up and Henrietta will get a treat.

Patsy Ann was a dog with a mission. She lived in Juneau, Alaska, from 1929 to 1942 and for many years she greeted every ship that came into Juneau's port. She was so faithful that in 1934 the mayor christened her the Official Greeter of Juneau, Alaska. The crews of ships began to expect her to be there and they were never disappointed. Her fame spread and she became the most photographed dog in the West since Rin Tin Tin.

Once, when erroneous information had been given out about which dock a ship was to pull into, people in the crowds waited patiently for their friends and relatives at the wrong dock. Patsy Ann, however, had an unerring sense as to where the ship would dock. She ignored the crowds and waited patiently in the right place and greeted the ship, alone.

The Lion and the Lamb

"And the lion shall lie down with the lamb" became more than just a prophecy at the Osaka Zoo in Japan. When still just a cub, a lion was rejected by his mother and raised on a diet of mostly vegetables and milk.

The lion cub proved so friendly, and was so lonely, that as an experiment the keepers decided to place a goat and later a sheep into the lion's cage.

What resulted was neither a feeding frenzy nor just three autonomous individuals who placidly ignored each other. Instead, the unlikely threesome became frolicsome playmates who clearly enjoyed each other's company. The goat would often butt at the lion, who would tussle happily with his friend.

The lion was much beloved by the zoo staff and learned to walk on a leash. Loneliness and the desire to make a friend proved stronger than the hunting instinct.

At Wildlife Images in Grants Pass, Oregon, a kitten wandered into the bear compound. The kitten must have been hungry because he approached the bear, asking for food. Dave Siddon, Sr., who was on duty at the time, thought the poor kitty was about to be torn apart by the bear. He watched in amazement as the bear offered some of his meat to the kitty.

The kitten and the bear became fast friends and the cat lived with the bear for three years. According to Vicky Reed of Wildlife Images, children in particular were delighted by the playfulness of the two. The cat liked to hide in the shrubbery and then jump out and attack the bear.

At night, the bear tucked the cat under his chin to sleep.

About forty years ago, anthropologists visiting Africa reported witnessing cooperation between some tribal people and a local lion population. They discovered that the local watering hole was shared—humans had it by day and lions at night. This truce was respected until cattlemen brought in cattle and used the waterhole at all hours to water their cattle. Once the unwritten law of sharing was violated, the lions responded by attacking the cattle.

Such stories remind us that there can be natural cooperation and intuitive balance between humans and the rest of the animal kingdom if we respect each other's ways.

It is not at all uncommon for domestic animals to assume caretaking roles for the offspring of other species.

As a girl, Diana Newbold lived in Northwest London with her cat Whiska. Whiska liked to sit atop the rabbit hutch and bat at the long ears of the male rabbits. However, when the baby rabbits wandered to the edge of the garden, Whiska took on the role of a sheepdog, crouching and herding the bunnies back to the safety of their own garden.

Many years ago a group of men who worked in a slaughterhouse stopped work for lunch. A lamb due to be slaughtered escaped from one of the pens and came over to join them in a friendly fashion. The slaughterhouse workers let the lamb nibble delicately on the lettuce in their sandwiches and petted the little creature as they all became acquainted. Proving, eloquently, that it is possible to kill only what one refuses to really know, the men finally sent the lamb away from the slaughterhouse. None of them had the heart to murder the creature.

As humans we often think that wild animals mate out of purely biological yearnings with little, if any, regard for the individual nature of their lover. Or we might think that only characteristics such as size or strength or physical stamina matter, because those traits would be valuable for defense and passing down to offspring. But just as humans often meet the partners of their friends and wonder, What does she see in him (or her)?, so there is a parallel in the animal kingdom.

Boo-Boo was young and very beautiful, by chimpanzee standards. The London Zoo administrators were excited about finding a mate for her and so, at great expense, had various virile young males shipped to the London Zoo. But Boo-Boo knew what she wanted and she spurned these males.

Finally, after Boo-Boo had rejected suitor after suitor, there were very few captive male chimpanzees left to try. A particularly unlikely prospect was an older chimp named Koko. He was balding, had a big paunch, and was generally unsightly, but he was shipped to the zoo for Boo-Boo's inspection anyway.

Boo-Boo must have seen an inner beauty that everyone else had missed, because one year later Boo-Boo gave birth to the first baby chimp ever to be born in the London Zoo.

When I first came across this story, my immediate thought was: Well, Boo-Boo was female. We females often fall in love with guys for their personalities or their character. Don't men pick their mates primarily on the basis of physical beauty, at least initially?

Not all males, it seems. At another zoo, a beautiful male parrot was introduced to a beautiful female. She was lovely to look at but apparently there was no chemistry between them because he refused to begin any of the mating rituals with her.

Another zoo contacted the first zoo saying they had a parrot so ugly that it was frightening the children who came by. She was old, and due to the stresses of captivity she had picked out all her feathers in frustration. Could the zoo just put her on a perch behind the two beautiful parrots so she could at least live? They agreed and she was shipped over.

The beautiful male parrot was immediately smitten by the featherless old female. He approached, he courted, they built a nest. Her feathers grew back and, although zoo officials considered her too old to be a mother, the two raised baby parrots—proving that beauty is more than feather-deep.

> *Man, when living, is soft and tender; when dead, he is hard and tough. All animals and plants when living are tender and delicate; when dead they become withered and dry. Therefore it is said: the hard and tough are parts of death; the soft and tender are parts of life.*
>
> —*Lao Tsu*

We are always fascinated by the many strange alliances and even love-matches between unusual animals that have been brought up together. Such a match was the friendship between Ginger the cat and Sam the seal.

This unlikely pair had both been reared by naturalists. When the two first met, Sam charged at Ginger to scare him. But Ginger was a brave cat who apparently believed in turning the other cheek: he simply stretched up to rub his face against the seal's long whiskers. This surprising affection completely demolished the seal's defenses.

The two became the best of pals. Sam's favorite game was to grab Ginger by the scruff of his neck and drag him to the center of a room. When Ginger tried to slink away, Sam would again grab the cat and push him back to the middle again. This game would continue for hours until Ginger tired of it and retreated to his box. If no plaintive seal coaxing sounds could tempt the cat out, Sam would content himself with pushing Ginger in the box around the floor with his nose. Ginger loved his seal and would eventually come out and rub his face against his friend's nose so the game could begin again.

Sam was so attached to Ginger that when the cat was stretched out by the warm stove, Sam would put up with the discomfort to lie by the hot stove next to his cat friend, even though his thick coat and heavy layer of insulating blubber were designed for diving into icy northern waters.

Friendship knows no boundaries. It seems that consciousness, love, and the ability to have fun and play together will often overcome all obstacles.

A tiny marmoset monkey and a chimpanzee were raised together. For hours they would play rambunctiously and get into the kind of trouble that only young monkeys and chimps can find.

One day, the naturalist who kept them thought that the chimp was getting too big for his tiny friend, and decided to separate them for fear of injury to the marmoset. He put the marmoset in a large birdcage which he hung from the ceiling and locked with an elaborate mechanism, and went off for the day.

By the time the man returned, the chimp had pulled a table under the cage, placed a chair on it so he could climb up to his friend, unlocked the elaborate mechanism, and the two buddies were happily wreaking havoc together again.

A Friend in Need

Della was mute and crippled and sharply felt the pain of being different from other youngsters. She loved her mixed-breed dog, Stubby, more than anything else.

Stubby was Della's greatest source of love and comfort in the world. He seemed to understand her even though she couldn't speak and he shared the little girl's troubles and pain. When Della's sadness became too great, she could always hold her little dog in her arms and know there was at least one being who completely understood her.

While Della and her family were traveling to Colorado Springs, Stubby was somehow lost along the border between Indiana and Illinois. The family car was traveling so fast that no one knew exactly where he had fallen out, so searching for him was extremely difficult. Newspaper ads proved futile.

To lose a beloved animal companion is always a terrible thing, but to lose one when you are a crippled and mute thirteen-year-old girl is doubly heartbreaking. Della's inability to speak meant that she could not even confide to anyone the pain of her loss. Stubby had been her truest friend. She felt life had dealt her a cruel blow.

But miracles do happen. One day, a year-and-a-half later and a thousand miles away, Della's grandfather saw a starving mongrel dog on the porch of their former home. Ragged and bruised from an incredible ordeal, the dog was so exhausted he could barely lift his head. His coat was dirty and his body distended from lack of food. Della's grandfather looked more closely and saw that his paws were bloody and raw as if he had walked a very, very long way.

It was indeed Stubby, who had been as faithful in looking for Della as she had been to him. He had always known he had to get back to his little girl. On the porch of the only home he had ever known, Stubby patiently waited for her. The two met in tearful reunion, and Stubby was nursed back to health. The little dog had walked a thousand miles to come back home to Della.

Love of animals is a universal impulse, a common ground on which all of us may meet. By loving and understanding animals, perhaps we humans shall come to understand each other.

—Dr. Louis J. Camuti

How is it pets can have a positive effect on our health?

For one thing, animal companionship helps us reduce stress. Animals allow us to be genuinely ourselves. They don't care about human distinctions of gender, creed, color, politics, and so on. Dr. Aline H. Kidd, professor of psychology at Mills College in Oakland, California, says that animals allow us to relate with a being who doesn't judge, doesn't argue back, doesn't have prejudices, biases, preconceptions, and definitions that don't agree with yours—and above all else, animals don't tell on you.

Dr. Michael Robins of the University of Minnesota found that pets play an important role in children's development, supplementing parental relationships. Pets provide unconditional affection which proved especially important in the case of disturbed teenagers. Ninety-seven percent of teenagers who owned pets reported that they loved them very much. Unlike friends or parents, who can cause enormous emotional suffering, teenagers found a reliable source of joy in their dogs and cats. In addition, the responsibility of caring for an animal encourages self-reliance and self-worth in children.

Andrew was very ill with AIDS. He was going home after a long stay in the hospital and knew that much of the life he had loved and appreciated would be different now. He was afraid that even his cat might be put off by him; after all, he had been away a long time, he smelled like a hospital, he had tubes in his body, and he was very weak and could no longer pick the cat up as he used to do. At night, Andrew had to hook himself up to a machine that supplied him with nutrients and medicine. Things would be very different now.

However, his cat Merlin was not at all put off by all this. In fact, Merlin seemed to understand. After Andrew had accomplished the necessary activities to get himself situated for the night, Merlin showed he, too, could adapt. Very consciously and deliberately the cat approached Andrew, carefully pawing his way up his friend's body, taking care not to jostle the tubes. He rested his head on Andrew's chest and then loudly purred his gratitude for Andrew's presence in his life again. The man's relief was tremendous. Some things can be counted upon.

Andrew says that AIDS has taken away most of his life from him, but even the ravages of a dread disease could not take away the love of his cat.

When animals were introduced into an institution for people with special needs, the caretakers noted that the patients who had a higher level of skills helped the more retarded patients to carefully and gently care for their new companion animals.

Thirty-one people with severe illnesses answered a questionnaire about the animals that had been given to them. Twenty-nine responded that their animals "helped me to laugh and maintain a sense of humor."

Jack Fyfe awakened to the awful realization that something was terribly wrong. His body didn't work as it had the day before. Then he realized what it was—a stroke! He tried to move but couldn't. What would become of the seventy-five-year-old man? People rarely visited; who could know what had happened to him?

Jack called to his sheepdog, Trixie. Trixie knew that something was wrong. Sheepdogs are an intelligent and resourceful breed, but what Trixie did went far beyond what anyone would expect from a dog. She found a towel, soaked it in her water bowl, and brought it back to Fyfe. He sucked the water from it. She stayed with her friend, lying faithfully at the foot of the bed. When the water in her bowl ran out, she got water from the lavatory bowl.

For nine days Trixie kept Jack alive. Finally, Jack's daughter became alarmed when he did not show up for a family dinner and found him. He had lost a lot of weight but was still alive, thanks to the ministrations of Trixie.

> *We cannot have two hearts, one for the animals, the other for man. In cruelty toward the former and cruelty to the latter this is no difference but in the victim.*
>
> *—Alphonse Marie-Louis d'Lamartine*

Rare Fellowship

Robert Franklin Leslie put up his lean-to and prepared to camp deep in the Canadian woods. The outdoorsman had spent thirty years hiking through the most isolated reaches of North America. When he climbed down to a nearby stream, he found himself with most unexpected company.

Thirty yards away stood a giant black bear. Robert knew he dare not make any fast movements. He also knew that the bear could kill him if he wanted to. Adventurer that he was, though, Robert decided to proceed calmly down to the stream and go fishing.

The bear followed.

Robert made slow, careful movements as he baited his hook and started to fish. The bear, who shared his interest in fishing, settled down just five feet away and watched attentively. Then Robert caught his first fish—fourteen inches—and tossed it to the bear.

The bear gulped it down and waited appreciatively until Robert caught another, which also went down the bear's gullet. And another. Robert fished and fed his catch to the bear until nightfall. Then he decided to head back to camp and cook up a couple of fish he had caught for himself.

The bear followed. Robert named him Bosco and spoke softly to him.

Robert made a campfire, ate his fish, and settled down to sleep in his sleeping bag. Bosco decided to settle down with his new fish-catching friend. Man and bear stretched out together under a tarp and enjoyed their quiet companionship. The smell of wet, furry, fish-sated bear filled Robert's nostrils. He decided it was rather a good smell.

Bosco got up in the night and attempted to scratch the lower part of his back, just above his tail—a difficult place to reach. He came to Robert, obviously asking for help. Inspecting the itchy spot carefully, Robert saw that it was full of fat ticks that were irritating the bear's flesh. The first tick he pulled out caused Bosco to roar so loudly it jolted the forest. For a moment, Robert thought the bear would kill him, but then Bosco seemed satisfied when he saw Robert pitch the tick into the flames. Extracting each tick from the inflamed flesh, Robert held it up to the bear to sniff and then cast it into the fire.

As they returned to sleep, Bosco was careful not to put the full 500 pounds of his weight against Robert. The bear's cold nose woke Robert up several times as he came and went during the night. The next morning, to Robert's extreme delight, Bosco chose to follow him as he hiked to the next destination.

Robert and Bosco got to know each other very well. They romped and wrestled. If Bosco's play became too rough, Robert had only to roll over and play dead to find Bosco whimpering and licking his face.

Robert chopped open a tree full of honeybees for Bosco, scratched him when asked, and fished for him when the bear desired. It was all a bear could hope for.

Robert studied Bosco's movements and vocabulary. Bosco would sometimes pick the man up in a heartfelt embrace that gave true meaning to the term "bear hug," and he would lick Robert's face.

Perhaps the best times of all were at night by the campfire, when Bosco would stare deeply into the man's eyes and commune as if the two were Man and Nature itself delving deep into one another. Bosco enjoyed this profound contemplation and, at the end of it, would often put a gargantuan paw upon the man's shoulder as if some meaningful point about the nature of life had been reached.

One day Bosco gave a signal to Robert not to move as they hiked through the woods. Suddenly they were surrounded by four bears. These bears were younger than Bosco, only two-year-olds, whereas Bosco was full-grown. He beat back each one in turn.

That night at camp, Bosco wanted to spend extra time gazing into Robert's eyes, although Robert didn't yet understand the reason for this. Perhaps the encounter earlier that day had reminded Bosco more fully that he was a bear, not a man. The next day, as they were hiking, Bosco ambled off in another direction. Robert didn't call to him. The bear didn't look back. Robert wrote, "He left behind a relationship I shall treasure."

> *One does not meet oneself until one catches the reflection from an eye other than human.*
>
> *—Dr. Loren Eiseley*

Scoffers object that animals who seem to love a person are really only in it for the little tidbits of food offered. Sometimes, perhaps—and yet this is often not the case. There are stories of animals in captivity who were fed by machine but still came to act affectionately toward the people who acted with kindness toward them.

Here is a wonderful story of the friendship between a usually reclusive bird, called the Lyre bird, and a woman named Edith Wilkinson who lived in a remote part of Australia.

James, as Edith called the Lyre bird, would visit her each day and sing and dance melodious songs of the wild. A great mimic, James would fill the air with forest sounds that he had heard in his travels and regale her with amazing displays of both dance and song.

At first Edith thought that her duty as audience to this performance was to bring the entertainer some offerings, and so she collected grubs and insects for him. As a lover who had given his love freely and was now offered payment, James was highly offended. The bird cawed loudly, ruffled his feathers, lifted his beautiful crest, and charged the food, sweeping it off the platform she had provided for him. After that, Edith decided that their relationship had a spiritual basis.

For years James courted and entertained her, absent only during migration or on the rare occasion when she told him she would not be home the next day—a communication he seemed to readily understand. The bird always resumed coming to the house once Edith had returned.

One time when Edith was ill, James appeared outside her home. Too nauseous to rise, she was thrilled to see his face in the window as he serenaded her with songs of good cheer and happiness. But, there being no tree outside her window, how had he managed to get up high enough to look in? Later investigation showed that the bird had moved garden debris under her windowsill so he could stand on it and see her. Naturalists interested in observing and studying the elusive Lyre bird documented these encounters. The relationship between Edith and her entertaining bird friend continued for many years.

I CANNOT TELL MY BIRD TO SING

I cannot hold the instrument
So the melody plays me,
I bought my bird a golden cage
Before I set her free.

I cannot prove the things I know
But knowing them is my proof
In this embrace, this endless love,
I cannot be aloof.
I cannot lock my garden door
Or keep myself apart
I know the world is under me
But I feel it in my heart.
I cannot sing the melody
With this poor instrument
I cannot speak the words I hear
Or say the thing I meant.
I cannot draw the circle right
It never comes out round
I cannot tell my bird to sing
But still I hear her sound.

—Jeffrey Armstrong

Parents often underestimate just how much an animal can mean to a child and vice-versa. Sometimes they learn the hard way.

After a vacation in the country, little Aurelie's parents were about to head home when they discovered that Scrooge, their daughter's cat, was missing. They searched but couldn't find the cat so they finally decided they had to leave without their daughter's beloved animal.

Aurelie cried bitterly all the way home. She was a child with special needs. When she was only four years old, she had taken a terrible fall that left her in a coma for a month. As a result, Aurelie was now was nearly blind, mute, and paralyzed down her left side. If ever a child needed someone to love her with the purity that only an animal can provide, it was Aurelie.

At home, Aurelie lost all interest in life. She stopped eating. At night she would pray for her Scrooge's return and stare at her cat's empty box in despair.

One day, nine months later, there was a scratching at the door.

Aurelie turned her wheelchair toward the door. Then two miracles occurred. The first miracle was the appearance of Scrooge at the door—feeble, miserable, and barely alive, but happy to be reunited with his little girl at last.

The other miracle came from within Aurelie herself. She spoke for the first time since the terrible accident had destroyed half her body and her power of speech. Astounding her parents. Aurelie cried out: "Scrooge! It's my Scrooge! He's come home!"

Aurelie's cat had walked 600 miles.

The miracles continued as Aurelie's speech improved. She eventually recovered most of her ability to walk. Needless to say, the chastened parents spared no expense for Scrooge's medical needs and Aurelie's fondest angel friend recovered as well.

Little children, never give
Pain to things that feel and live;
Let the gentle robin come
For the crumbs you save at home:
As his meat you throw along
He'll repay you with a song.
Never hurt the timid hare
Peeping from her green grass lair.
Let her come and sport and play
On the lawn at close of day.

The little lark goes soaring high
To the bright windows of the sky.
Singing as if twere always spring,
And fluttering on an untired wing
Oh! Let him sing his happy song.
Nor do these gentle creatures wrong.
—*Anonymous*

This story could be called the bovine version of *The Fugitive.*

Missy Cow was being washed and prepared for slaughter along with two other yearling calves near Leavenworth, Louisiana. Perhaps sensing the fate that awaited them, she and her two companions suddenly bolted for freedom.

The slaughterers grabbed their guns and hunted down the two calves. But Missy Cow headed for the Cascade mountains and escaped!

Thus began Missy Cow's great adventure, which was to become almost legendary in the state of Washington. The cow's first big challenges were to find a way to live through harsh winters, fight off coyotes, and find food. For seven years, she survived.

After she had been wild for several years, Missy Cow made friends with a dog named Keesha. Keesha's humans appreciated the bizarre friendship and fed Missy during the worst parts of the winter.

After a couple of years, Keesha's humans realized they could not afford the cow's upkeep any longer. Not knowing what else to do, they decided to let the slaughterers finally have her.

As the farmer's wife wept openly for their betrayal of the poor animal, the men took aim. But all three of their bullets just bounced off Missy Cow's head and neck. It seemed like a miracle, but Missy's seven years of hard living had made her hide so tough she was impervious to .22 caliber bullets.

While the men went back to get bigger weapons, the news of Missy Cow's death-defying story reached Diane and Gene Chantos, a couple who rescue wild burros. So they decided to drug and rescue Missy Cow, too. When the men returned with their high-powered rifles, Missy Cow was nowhere to be seen.

Said the rancher who had fed her through the terrible winters, "My wife said God protected her. Well, who am I to say He didn't?"

In separate and unrelated incidents, two Golden Retrievers, Zeke and Doc, each rescued a man who would have died without their aid. Zeke saved Lester Needham, who had been missing for three days at Yosemite National Park after he fell into a chasm and was too injured to move. Doc saved Jeff Eckland, who was buried under four feet of snow in an avalanche. The two dog heroes are father and son.

Our Common Condition

Compassion is the highest form of human existence.
—Fyodor Dostoyevsky

I was talking with the receptionist at my veterinarian's office about a necessary but painful procedure for one of my two cats. Though sweet and well-meaning, the receptionist said, "It will be all right. Animals don't feel pain in the same way we do."

I wondered on whose authority she got this information. She didn't claim to be an animal psychic so she wasn't getting it from the animal, that was for sure.

"No," I told her, "actually animals might feel pain even more deeply than we do."

How do we know this isn't true? For one thing, an animal's survival is based on sharper senses. Their sense of smell may be hundreds of times greater than ours, as well as their sense of hearing. Can any human claim to have the eyesight of an eagle? I have noticed that my companion animals' nerve endings react instantly to the gentlest caress, and that if I should accidentally step on a tail they react as I would if a giant stepped on my foot. Dr. Louis J. Camuti states, "Never believe that animals suffer less than humans. Pain is the same for them that it is for us. Even worse, because they cannot help themselves."

Therefore it seems reasonable to conclude that an animal's awareness of physical sensation is at least the same as mine or deeper.

Animals react just as strongly to emotional pain. Dogs have been known to lie down upon their human's grave and refuse food until they starve to death. Pat Derby, founder of Performing Animal Welfare Society (PAWS), tells the story of a South American cougar left with her after he was abandoned by a wealthy couple who had cared for the cougar for the first five years of his life. The couple, who claimed to love the animal and who had enjoyed having an exotic pet, simply parked their cougar on Ms. Derby's floor and departed.

Pat and her husband endeavored to make the cougar feel comfortable. They arranged their schedules so they could spend a lot of time with him. They spent twelve hours cuddling him and trying to get him to eat something. But he never moved or took his eyes off the door where the only two friends he had had since he was a cub departed.

After a few days Pat called the vet, who told her that the cougar was simply willing himself out of existence. The wealthy couple declined to see him even when told of his dire condition. Within a week the heartbroken cougar died.

Pat said that hard as they tried to make him happy, "I don't think he ever really saw us. Only two people had existed for that cougar. And now they were gone."

This is one great reason why you should never adopt an animal unless you are willing to make a lifetime commitment to it. Animals are not separate and independent beings. They care, they connect, they love. And sometimes they can love too much.

> *Men have forgotten this truth, but you must not forget it. You remain responsible forever for what you have tamed.*
>
> —*Antoine de Saint-Exupéry,* The Little Prince

The argument is often made to people who care about animals that they shouldn't waste their feelings on animals when there are human beings who are hurting. When asked why he spent so much time and money talking about kindness to animals when there is so much cruelty to humanity, George Angell would answer, "I am working at the roots."

This argument of the priority of human rights over animal rights is often used in the case of children: "How dare you make so much effort to spare animal suffering, when so many of our children are starving or abused?"

But the two issues are absolutely interconnected. This interconnection has been legally recognized for a long time. The first court ruling in the United States for a child who was being abused could not have happened if animal rights had not first become law.

In 1874, a little girl named Mary Ellen was brought before the courts to seek protection. She was filthy, wore rags, and had wounds and scars all over her body. Her stepmother beat her regularly.

The judge, horrified by what he saw before him, searched in vain for a law that could have her removed from her family and put the stepmother in jail. But in 1874 there were not yet any laws on the books protecting the rights of children. Fortunately, animal rights legislation had just been passed.

The judge ruled that whatever else the little girl was, she was also an animal, albeit a human one. Therefore, he reasoned, she was entitled to full protection, as an animal, under the law. She was taken from that family and her stepmother was sentenced to the maximum time in jail for animal cruelty.

Soon after, Henry Bergh, the founder of the SPCA, started an organization for the prevention of cruelty to children as well.

> *So often when you start talking about kindness to animals someone comments that starving and mistreated children should come first. The issue can't be divided like that. It isn't a choice between animals and children. It's our duty to care for both. Kindness is the important thing. Kids and animals are our responsibility.*
>
> *—Minnie Pearl*

How often animal lovers have been scolded for admiring the spiritual grace of animals: "You are projecting human qualities on mere animals," people say. "Animals are not like us; they're not human." Yet American Indian spokesperson Ohiyesan stated that it has not always been this way. "The first American mingled with his pride a singular humility. Spiritual arrogance was foreign to his nature and teaching. He never claimed that the power of articulate speech was proof of superiority over the dumb creation...."

In his fascinating book *Divorce Among the Gulls,* William Jordan discusses his views on anthropomorphism after spending three years watching cockroaches. He finally decided that even *their* behavior was a little too human for his comfort.

> To think that there might be some commonality in the workings of the animal and the human mind was ridiculed by the university crowd, who called it by the grandiloquent term "anthropomorphism." The term meant blasphemy: Thou shalt not read the motives of man into the dim-witted brains of vermin. The very premise was considered arrogant and self-aggrandizing, a kind of humanistic public relations job. One cannot use the term anthropomorphism without presuming a grand rift, a kind of holy, unbridgeable chasm between the minds of all other animals and

the hairless ape. Humans march to the tune of the rational mind, while abysmal creatures like the cockroaches follow blind urges and deaf desires called instincts.

Yet I couldn't dismiss what I had seen. In some nagging way the notion of anthropomorphism went against biology. What if the intellectual establishment had it backward? What if, instead of imputing human thought to the animal mind, we should impute animal workings to the human mind? If indeed we have evolved from animals, what was the human mind but an extension of the animal's urges?

Ask the experimenters why they experiment upon animals, and the answer is: Because the animals are like us. Ask the experimenters why it is morally okay to experiment on animals, and the answer is: Because the animals are not like us.

—Professor Charles R. Magel

The Great Experiment

The most inspiring acts of kindness often take place under the most awful circumstances. Just as a small kindness can take on massive proportions in a prison or concentration camp, so a gentle reassurance can mean much more than it ordinarily would under adverse circumstances.

Hans Reusch tells this story of two dogs acting with compassion toward each other in a laboratory where experiments were performed upon animals.

A small dog had been partially paralyzed in an experiment that took away its ability to use its back legs. He was left to suffer alone on the floor, ignored by the researchers.

But then he spotted another little dog in the room. Despite his painfully wounded body, the partially paralyzed dog started to drag himself over to the other dog. This dog could not see his eager visitor because he had been blinded in another experiment and his eyes were putrefying.

A research assistant happened to turn, at that moment, and watched as the crippled little dog licked the blinded one giving him what comfort and love he could. "That pathetic gesture of mutual sympathy," the assistant later wrote, "put the human race to shame."

> *Wherever the art of medicine is loved, there is also a love of humanity.*
>
> —*Hippocrates*

In another case, researchers discovered that when two capuchin monkeys are put in a cage with a wire mesh between them, the one that is fed will push food through the mesh to the other. Perhaps what we most need to improve our treatment of animals is a little common sense. As Oliver Wendell Holmes said, "Science is a first-rate piece of furniture for man's upper chamber if he has common sense on the ground floor."

FRAGILE TRUST

It is a fragile trust that we are given
All creatures' fate in our dominion
To hold or hurt
to harm or cherish
To nurture life
or let it perish.
—*Stephanie Laland*

Probably no word in the English language is more overburdened than the word "dominion." Sometimes I think it must surely feel crushed under the weight of all the rationalizations it is made to carry. Every horse that was ever beaten, every puppy hurt—the poor word has had to prop up a world of justifications for cruelty on its sagging shoulders. All because in the King James Version of the Bible we are told that we have been given "dominion over the animals." And so a few thousand years of cruelty rides on the back of one word.

Of course, there are those who believe that the original Greek word translated as "dominion" had a gentler coloring to it. They point out that dominion does not necessarily mean "domination," which has a master-slave connotation.

It is interesting to note that the men making the most out of the Biblical dominion passage always seem to be those who blithely ignore all the rest. By their behavior at least, many who have justified their actions by this passage seem not to have taken any other part of the Bible seriously, particularly the injunction to "love thy neighbor as thyself."

I interpret the word dominion to mean something closer to "care." After all, if one goes to the movies and leaves one's children with the babysitter, the children are always told to mind the babysitter. In a sense, the babysitter has dominion over the children. But if parents should come home and find the babysitter has hurt their children, we recognize that would be wrong, a misuse of the babysitter's dominion. The children's sheer innocence and inability to defend themselves would make it doubly wrong.

So it is with animals.

DOMINION

Let us update dominion
From antiquated opinion
A gullible minion
Of sanctioned control.

One must look within one
Your heart knows the wisdom
Not harming God's creatures
Is written in Soul.
—*Stephanie Laland*

Most experiments conducted on live animals are cruel, heartless, repetitive, and even unnecessary to human healing. One hundred million animals die each year in useless experimentation. The cures for polio, smallpox, insulin treatment for diabetes, and the major diseases from which humankind used to suffer have been discovered without the use of animal tests. That animals are tortured to death to test everything from oven cleaner to detergent to cosmetics is, to me, an obscenity against nature and God.

Once one accepts that there is something essentially different between a dog chasing after a stick and a stone plummeting to earth, then one will have a hard time in morally justifying, for example, why a healthy dog is given lung cancer with tobacco smoke in order to prove something to a suicidal human who smokes forty cigarettes a day.

—Stanley Godlovitch

Still, I must admit that there is one experiment I like.

Investigators at Ohio State University were researching the effects of diet-induced atherosclerosis in rabbits. All was going well—which, in this case, means the rabbits were getting diseased—in all study groups except one. The researchers were able to induce atherosclerosis by feeding the rabbits high-cholesterol diets, but one group inexplicably had 60 percent less atherosclerosis. The experimenters were baffled and tried to find the responsible factor. Nothing they tried proved to be it—not diet, not room temperature, not anything they could change and measure.

Finally they discovered that the particular researcher in charge of that group really liked rabbits. He would talk to them, pet them, give them lots of love. So the experimenters staged other experiments in which control rabbits were ignored while other groups were cuddled and talked to while all other variables were kept the same. Sure enough, in every case the rabbits that had been loved had at least 60 percent less incidence of atherosclerosis than those that were not shown affection. Atherosclerosis, by the way, is statistically the disease that kills most Americans. And, at the time this study was done, there was no drug against this disease that was effective in 60 percent of the cases.

I was so affected by reading about this experiment that I looked up the original study to see if this could possibly be true. Indeed, it is!

The lesson here is that love is powerful medicine. Love and compassion are necessary to us all and can even affect research. Caring and loving feelings can reach out and affect diseases and create healing—even when you are experimentally trying to induce an illness.

And so, I cannot be said to be against *all* animal experimentation, although I do not approve of the part about putting rabbits in cages and feeding them bad food. Instead of torturing animals in laboratories, as we are now, perhaps we will find the cures for the diseases we fear, by instead *loving* all beings. Perhaps all of us who truly care for other creatures can begin the greater experiment of loving—loving all life, at all times, in all the ways we can.

Let us begin the experiment of not turning our backs on helpless creatures who have hearts and souls because we hope that medical researchers might fix some ailment through cruel experiments that we have collectively decided we are willing to allow.

It has been said that all things are connected. It has been said that man did not weave the web of life—he is merely a strand in it. And that whatever he does to the web he does to himself.

With the advent of the new millennium, all humanity anticipcates a new stage in human evolution. The old worldview separates us from and objectifies nature. It sees animals merely as things to be dominated. This is a schism in our collective soul. As part of the global spiritual transformation, we are coming to the momentous realization of the underlying force in every living creature. If the animal kingdom is to survive into the new millennium, we must build an ark of caring and compassion to protect and nurture it. We must begin to nurture Nature herself.

Life itself is a great experiment. We have always said, because we have always intuitively known, that love is the answer. Let us begin by loving all the creatures from the least to the greatest and by protecting them all.

Animal angels? It is love which makes angels of us all.

The Ark of the New Millennium

The single most significant thing I have learned in the study of the animal kingdom is that compassion and kindness are universal qualities, not limited to our own species. The amazing actions of animal angels have forced me to expand my view of the world. I find it deeply inspiring to know that dogs understand and show compassion to their human friends when a relative has been killed, that a family cat might bite the leg of a mother whose baby is crying to warn her not to hurt it, that a bigger bird would lift its wing to shelter a tiny lonely one, and that a rat would alert a sleeping little boy to danger instead of just saving its own life.

These stories reveal a world that is kinder and more beautiful than I once believed. This is good news, for studies among humans have indicated that just watching an act of kindness predisposes the observers to perform an act of kindness themselves.

We may be already living in a better world than we thought we were if we take the time to really look.

Humans have a habit of characterizing our regrettable inhumanity to each other as "animality," when in fact, animals are rarely cruel. Within the framework of the great recycling plant called Earth, predators, while they kill for food, rarely do this out of any sense of cruelty. When sated, they cease. They do not massacre, wage war, clearcut forests, or devise artificial compounds that threaten to destroy us all.

In fact, as we have seen, kindness is common among all nonhuman species. Animals may even be as humane as any being on the planet.

The actions of animals speak louder than any words and must surely sweep aside all human doubt that we are separate from or morally superior to animals.

It wasn't too long ago when society felt it was perfectly alright for certain people to be used in ways that we would never countenance now, simply because of their skin color. As the human race evolves we are including different parts of our own kind into the family. Now we are beginning to reach out and realize that there is a divine spark within *all* life. We have to keep letting go of our old boundaries by acknowledging that if a being is capable of compassion, feeling, and caring, then it must be treated with respect.

Today it is more critical than ever for human beings to realize the sacredness of all life. The issue is as important as whether or not we will be permitted to go on living on this planet, or perhaps simply whether or not we will be able to endure it here. The answer depends on whether we, the human race, treasure the beauty and preciousness of life, of love, not only in ourselves and our fellow human beings, but within all our animal brethren as well.

> *The human spirit is not dead. It lives on in secret. I have come to believe that compassion, in which all ethics*

must take root, can only attain its full breadth and depth if it embraces all living creatures and does not limit itself to mankind.

—Dr. Albert Schweitzer

As we approach the new millennium, the international Random Acts of Kindness movement continues to inspire people around the world, now embracing the animal kingdom as well. All over the world one can feel the quickening of consciousness as we approach the next step in human evolution—the realization of the underlying spiritual force in every being and in every molecule of existence.

The old concept of a mechanical, objective world is dying. We are being forced to recognize that nature and animals are not things to be dominated, but dear fellow travelers on the path of life. As the world rouses itself to heal our damaged environment, we humans are trying as never before to enter into communion with the intelligent, compassionate beings with whom we share this planet. This shift in consciousness is evidenced by the international movement to protect endangered wild species and abolish all suffering.

Be inspired with the belief that life is a great and noble calling; not a mean and groveling thing that we are to shuffle through as we can, but an elevated and lofty destiny.

—William E. Gladstone

It is entirely appropriate that kindness to our animal brethren should be enacted into law. There is no reason why governments shouldn't reflect the kindness in the hearts of the people who comprise them. But until we honor the spiritual connection between all beings, even law cannot halt the destruction of the environment or the meaningless testing of household cleansers on the helpless, sensitive animals who trustingly look to our species for love and guidance.

We are discovering that all things in nature really are indissolubly connected, and that we humans did not weave the web of life, but are merely strands of it. The environmental catastrophes of the last century have taught us the hard way that whatever we do to the web we do to ourselves.

In their worship of the machine, many Americans have settled for something less than a full life, something that

is hardly even a tenth of a life, or a hundredth of a life. They have confused progress with mechanization.
—Lewis Mumford

Certainly our attitude toward the planet changes our perception of the planet itself. Consider Chief Luther Standing Bear's description of the Indians' perception of Earth before the white men came:

"We did not think of the great open plains, the beautiful rolling hills, and winding streams with tangled growth as wild. Only to the white man was nature a wilderness, and only to him was the land infested with wild animals and savage people.

"To us, it was tame. Earth was bountiful and we were surrounded by the blessings of the Great Mystery. Not until the hairy man from the east came was it wild for us."

There is a spirit that runs through all of us, an impulse to goodness that heightens and uplifts all life. When one being acts with kindness and love and helpfulness to another, it uplifts us all.

These, then, are the precious strands in the web of life: love, helpfulness, kindness. These strands of compassion hold the world together, for without them we experience only separateness and pain. Life is inherently spiritual and moves us to respect all life forms—including those unlike our own. From earthworms to puppies, from kittens to rats, life contains kindness within its very essence.

What a beautiful, wonderful world.

Animals share with us the privilege of having a soul.
—Pythagoras

Sacred earth, sacred mother, sacred creature
Upon her breast
Each knowing his own life
Each part of the divine
Eternal partners in life's eternal quest.
Within each being; a spark of soul
Each form some aspect of the divine.
Forgive us for not seeing how precious
Each one of you are;
We can barely see it in each other
Or ourselves.
Trees, Earth, Air, Water, Animals
Why is it only as we lose you we see
How great is the offering of you to our planet?
Why must mankind desolate you to realize
How precious the gift?

—Stephanie Laland

BIBLIOGRAPHY

Amon, Alice. *The Earth is Sore*. New York: Atheneum, 1981.

Amory, Cleveland. *Animail*. New York: Windmill Books, Inc. & E. P. Dutton & Co., Inc., 1976.

____. *The Cat Who Came for Christmas*. Boston, MA: G. K. Hall & Co., 1987.

Bardens, Dennis. *Psychic Animals*. New York: Henry Holt and Company, 1987.

Barry, J. "General and Comparative Study of the Psychokinetic Effect on a Fungus Culture," *Journal of Parapsychology*, 1968.

Berlitz, Charles. *Charles Berlitz's World of the Incredible But True*. New York: Ballantine Books, 1991.

Boone, J. Allen. *Kinship With All Life*. New York: Harper & Row, 1954.

Carson, Gerald. *Men, Beasts, and Gods: A History of Cruelty and Kindness to Animals*. New York: Charles Scribner's Sons, 1972.

Cavalieri, Paola, and Peter Singer. *The Great Ape Project*. New York: St. Martin's Press, 1993.

Coudert, Jo. "A Shepherd's Healing Power," *Reader's Digest,* December 1996.

Darwin, Charles. *The Descent of Man*. New York: D. Appleton & Co., 1896.

Delehanty, Hugh, ed. "What Animals Could Tell Us," special issue of *Utne Reader*, March-April 1998.

Derby, Pat, and Peter Beagle. *The Lady and Her Tiger*. New York: Ballantine Books, 1977.

de Waal, Frans. *Good Natured: The Origins of Right and Wrong in Humans and Other Animals*. Cambridge, MA: Harvard University Press, 1996.

Dickson, Lovat. *Wilderness Man*. New York: Atheneum, 1973.

Dossey, Larry. *Healing Words*. San Francisco: Harper, 1993.

____. *Space, Time, and Medicine*. New York: Shambhala/Random House, 1982.

Ehmann, James, and David Gucwa. *To Whom it May Concern: An Investigation of the Art of Elephants*. New York: Penguin Books, 1985.

Evans, Rose. *Friends of All Creatures*. San Francisco: Sea Fog Press, Inc., 1984.

Fogle, Bruce. *Pets and Their People*. New York: Viking Press, 1984.

Fortean Times. London: John Brown Publishing LTD. www.forteantimes.com.

Fox, Michael F. *Returning to Eden*. New York: Viking Press, 1980.

Freshel, M. R. L. *Selections from Three Essays by Richard Wagner With Comment*. South Lancaster, MA: The Millennium Guild, 1954.

Gaddis, Vincent, and Margaret Gaddis. *The Strange World of Animals and Pets*. Stamford, CT: Cowles Book Company, Inc., 1970.
Greene, Lorne. *The Lorne Greene Book of Remarkable Animals*. New York: Simon & Schuster, 1980.
Hancock, Lyn. *There's a Seal in my Sleeping Bag*. New York: Knopf, 1972.
Helfer, Ralph. *Beauty of the Beasts*. Los Angeles: Jeremy P. Tarcher, 1990.
"Jake the Cat." Reuters World Service.
Jordan, William. *Divorce Among the Gulls*. New York: Farrar Straus Giroux and North Point Press, 1991.
Knapp, Don. CNN Online, April 10, 1997.
Kowalski, Gary A. *The Souls of Animals*. Walpole, NH: Stillpoint Publishing, 1991.
Lauber, Patricia. *Earthworms, Underground Farmers*. Champaign, IL: Garrard, 1976.
McLuhan, T. C. *Touch the Earth*. New York: Promontory, 1989.
Montgomery, Sy. *Walking with the Great Apes*. New York: Houghton Mifflin, 1991.
Moore, J. Howard. *The Universal Kinship*. Chicago, IL: Charles H. Kerr & Company, 1906.
Morris, Desmond. *Animal Days*. New York: William Morrow and Company, 1980.
Nerem, Levesque, Cornhill. "Social Environment as a Factor in Diet-Induced Atherosclerosis," *Science* 208, June 1980.
Newkirk, Ingrid, ed. "Missy's Great Escape," *PETA Catalog*, Spring 1998.

Patterson, Francine. *Koko's Kitten*. New York: Scholastic, Inc. and The Gorilla Foundation, 1985.

Pratt, Ambrose. *James, the Lyre Bird*. New York: Angus & Robertson, 1955.

Rasa, Anne. *Mongoose Watch*. New York: Anchor Press and Doubleday & Co., 1985.

Reader's Digest Association, Inc. *Animals Can Be Almost Human*. 1979.

____. *Animals You Will Never Forget*. 1969.

Regenstein, Louis. *The Politics of Extinction*. New York: MacMillan, 1975.

Robbins, John. *Diet for a New America*. Walpole, NH: Stillpoint Publishing, 1987.

Roberts, Yvonne. *Animal Heroes*. London: Pelham, 1990.

Romanes, George John. *Animal Intelligence*. Washington, DC: University Publications of America, 1977.

Ross, Andrew. "Merlin's Love," *Cat Fancy Magazine*, October 1991.

Ruesch, Hans. *Slaughter of the Innocent*. New York: Bantam Books, 1978.

"Simian-Pure Assistant," *Life Magazine*, 1955.

Singer Peter. *Animal Liberation*. New York: Avon Books, 1975.

Steiger, Brad. *Cats Incredible*. New York: Penguin, 1994.

Steiger, Brad, and Sherry Hansen Steiger. *More Strange Powers of Pets*. New York: Donald I. Fine, 1994.

____. *Strange Powers of Pets*. New York: Donald I. Fine, 1992.

"The Story of Bella." *Praetoria News*, November 11, 1997.

"The Story of Patsy Ann," www.patsyann.com.

Surran, Mary Anne. "Letter to the Editor," *People Magazine,* August 4, 1997.
Tedder, W., and M. Monty. "Exploration of Long-Distance PK," *Research in Parapsychology 1980*. Metuchen, NJ: Scarecrow Press, 1981.
Thomas, Elizabeth Marshall. *The Tribe of Tiger*. New York: Simon & Schuster, 1994.
"The Violence Connection," pamphlet published by the Doris Day Animal League, 1997.
Wallechinsky, David, and Irving Wallace. *The People's Almanac Presents the Book of Lists: The 90's Edition*. Boston: Little, Brown & Company, 1993.
Wallechinsky, David, Irving Wallace, and Amy Wallace. *The People's Almanac Presents the Book of Lists*. New York: Morrow, 1977.
Watson, E. L. G. *Animals in Splendour*. Camp Hill, PA: Horizon Press, 1967.
Wels, Byron G. *Animal Heroes: Stories of Courageous Family Pets and Animals in the Wild.* New York: Macmillan, 1979.
White, Betty, with Thomas J. Watson. *Betty White's Pet-Love*. New York: William Morrow and Company, 1983.
Wood, Gerald L. *Guinness Book of Pet Records*. London: Guinness Superlatives Ltd., 1984.
Wylder, James. *Psychic Pets*. New York: Stonehill Publishing Co., 1978
Wynne-Tyson, Jon. *The Extended Circle*. New York: Paragon House, 1985
Yerkes, Robert M., and Ada W. Yerkes. *The Great Apes: A Study of Anthropoid Life*. New Haven, CT: Yale University Press, 1929.

RESOURCE GUIDE

Animals Agenda
P.O. Box 25744
3201 Elliot Street
Baltimore, MD 21224

Animal Legal Defense Fund
1363 Lincoln Ave
San Raphael, CA 94901
(415) 459-0885

Association of Veterinarians for Animal Rights
P.O. Box 208
Davis, CA 95617-0208
(916) 759-8106

The Cousteau Society
930 West 21st Street
Norfolk, VA 23517
(804) 627-1144

Doris Day Animal League
900 2nd Street N.E., Suite 303
Washington DC 20002
(202) 842-3325

ETHIC (Every Tag Helps Indentification Campaign)
P.O. Box 1234
Capitola, CA 95010
(408) 423-1156

Fund for Animals
200 West 57th Street
New York, NY 10019
(212) 246-2096

The Gorilla Foundation
P.O. Box 620640
Woodside, Ca 94062
(800) 634-6273
(415) 851-8505
www.gorilla.org

Greenpeace
1611 Connecticut Ave. N.W.
Washington DC 20016
(800) 916-1616

In Defense of Animals
131 Camino Alto, Suite E
Mill Valley, CA 94941
(415) 388-9641

Last Chance for Animals
8033 Sunset Blvd, Suite 35
Los Angeles, CA 90046.
(310) 271-6096

Last Chance for Animals
18653 Ventura Blvd, Suite 356
Tarzana, CA 91356
(818) 760-2075

People for the Ethical Treatment of Animals (PETA)
P.O. Box 42516
Washington, DC 20015
(301) 770-7444

Performing Animal Welfare Society
P.O. Box 842
Galt, CA 95632

Physicians Committee for Responsible Medicine
P.O. Box 6322
Washington, DC 20015
(202) 686-2210

Tony La Russa's Animal Rescue Foundation (ARF)
P.O. Box 6146
Concord, CA 94524
(510) 685-1ARF

If you are interested in persuading landlords to change their policies regarding pets, contact The Open Door Campaign at the San Francisco SPCA, 2500 16th Street, San Francisco, CA 94103. This organization has successfully found ways to open the doors of closed complexes to animal friends.

Do you have a story to share about an animal that you know? If so, please write me at:

Stephanie Laland
c/o Conari Press
2550 Ninth Street, Suite 101
Berkeley, CA 94710

ABOUT THE AUTHOR

A well-known speaker on animal-related issues, **Stephanie Laland** is the author of *Peaceful Kingdom: Random Acts of Kindness by Animals* and *51 Ways to Entertain Your Housecat While You're Out.* A workshop leader for people wishing to increase their connection to animals, she and her husband and many animal friends live in Felton, California.

CONARI PRESS, established in 1987, publishes books on topics ranging from psychology, spirituality, and women's history to sexuality, parenting, and personal growth. Our main goal is to publish quality books that will make a difference in people's lives—both how we feel about ourselves and how we relate to one another.

Our readers are our most important resource, and we value your input, suggestions, and ideas. We'd love to hear from you—after all, we are publishing books for you!

To request our latest book catalog, or to be added to our mailing list, please contact:

CONARI PRESS

2550 Ninth Street, Suite 101
Berkeley, California 94710-2551
800-685-9595 • fax: 510-649-7190
e-mail: Conaripub@aol.com
Website: http://www.readersNdex.com/conari/